# Advances in Spatial Science



Springer
*Berlin*
*Heidelberg*
*New York*
*Barcelona*
*Budapest*
*Hong Kong*
*London*
*Milan*
*Paris*
*Tokyo*

## Titles in the Series

*Cristoforo S. Bertuglia, Manfred M. Fischer*
*and Giorgio Preto (Eds.)*
Technological Change, Economic Development and Space
XVI, 354 pages. 1995

Harry Coccossis
Peter Nijkamp (Eds.)

# Overcoming Isolation

## Information and Transportation Networks in Development Strategies for Peripheral Areas

With 64 Figures

Springer

Professor Dr. Harry Coccossis
University of the Aegean
Department of Environmental Studies
17 Karantoni str.
GR-81100 Mytilini, Greece

Professor Dr. Peter Nijkamp
Free University
Department of Economics
De Boelelaan 1105
NL-1081 HV Amsterdam
The Netherlands

Cataloging-in-Publication Data applied for

Die Deutsche Bibliothek - CIP-Einheitsaufnahme

**Overcoming isolation** : information and transportation networks in development strategies for peripheral areas / Harry Coccossis ; Peter Nijkamp (ed.). - Berlin ; Heidelberg ; New York ; Barcelona ; Budapest ; Hong Kong ; London ; Milan ; Paris ; Tokyo : Springer, 1995
(Advances in spatial science)
ISBN-13:978-3-642-79829-0 e-ISBN-13:978-3-642-79827-6
DOI: 10.1007/978-3-642-79827-6

NE: Coccossis, Harry [Hrsg.]

ISBN-13:978-3-642-79829-0


Softcover reprint of the hardcover 1st edition 1995



SPIN 10488593 42/2202 - 5 4 3 2 1 0 - Printed on acid-free paper

# Preface

As European countries pursue a common effort towards establishing a European Union, various isolated - and consequently disadvantaged - regions are likely to face increasing competitive pressures due to their peripheral location. To assist such areas, regional, national and supranational bodies put much effort into developing transport and communication networks and linkages in order to ensure that such less favoured areas are better integrated in the broader European social and economic development process.

This book addresses the issue of lagging development in various - mainly central and southern - European regions which are in a disadvantageous position as a result of their isolated location. The persisting problems of social and economic development in several European Union areas (e.g. islands, mountains, border areas) has turned the attention of policy-makers to the critical importance of transport and (tele)communication linkages.

The purpose of the book is to bring into perspective the role of transport and communications in regional policy for peripheral areas. This subject is currently of high priority, since the European Union through the Structural Funds interventions (i.e. the Community Support Frameworks) and the new Cohesion Fund relies heavily on transport and communication infrastructure investments to assist areas which are at a disadvantage due to their peripheral location and isolation. Furthermore, as the Union considers enlargement, some of these issues might be of wide European interest.

Most of these interventions are based on several assumptions about the potential contribution of such projects/programmes to economic and social development, derived mainly from the experience of mainly western industrialized countries operating in a quite different context from that of southern Europe in the nineties. These differences stem from the state of social and economic development and the processes of transformation, institutional changes in the broader international context (i.e. the Single European Space), the prospects of technological change in transport and communications and their potential impacts on the location of people and their activities. Therefore, policy responses might be quite different from those envisaged and intended, in a positive or negative sense. Special issues involved deal with missing linkages, barriers, accessibility and peripherality/isolation. New approaches might be required which call for better integration of economic development policy in physical planning taking into consideration the new spatial dynamics involved.

This book offers many new insights for researchers, academics, professionals and policy-makers in the fields of transport, telecommunications, regional development and economic development at the regional, national and supranational level. It is organized in two parts:

- The first part is intended to reflect on the general issues involved in transport, communications and regional development in peripheral areas, with a specific view on policy relevance.
- The second part aims to offer various empirical insights and examples of current advanced research in the area of policy analysis in the above issues. It also attempts to highlight some of these issues by focusing in particular on key problems faced by peripheral - mainly southern - areas of the European Union.

The editors are grateful to all authors who were prepared to release much strategic information with a high policy relevance for the purpose of this publication. This publication would not have come into being without the continuous support of Gwyneth Moore (Contact Europe) who demonstrated her master skills in text editing for this book. Finally, the editors wish to mention the professional drive of Marianne Bopp (Springer-Verlag) who realized that isolation can only be overcome by close cooperation.

Athens, Harry Coccossis
March 1995 Amsterdam, Peter Nijkamp

# Contents

# Part A

# Isolation and Peripherality: The Role of Networks, Borders and Barriers

# 1 Borders and Barriers in the New Europe: Impediments and Potentials of New Network Configurations

P. Nijkamp

## 1.1 New Megatrends: New Opportunities?

Spatial-economic connectivity and changes in industrial organization have far reaching consequences for the competitive profile and position of all regions in a network economy. In particular, geographically isolated regions have expressed a concern that they may find themselves positioned outside current mainstream industrial developments. The industrial-economic systems of our world are indeed rapidly changing, at all geographical levels. The traditional large scale production plant is gradually losing its relevance. In a post-fordist economy we observe much more emphasis on flexible entrepreneurial behaviour based on *lean production*. Modern industrial production is characterized by both *specialization* and *globalization*, in which the modern component industry and industrial assembly play an important role (see Lagendijk 1993). An example may clarify the above point. The Swedish Volvo company is producing its Volvo 480 in the Netherlands. The components of this car originate from different countries: 28% from Germany, 26% from France, 22% from Belgium, 12% from the Netherlands and 12% from remaining countries (e.g. Sweden, Denmark, Finland, the UK, Switzerland, Spain, Japan, the USA, Canada, Brazil and Australia).

Flexible production in a globalizing economy means that the *national identity* of products as 'national flagships' is disappearing. Global alliances and partnerships generate some sort of a 'globalisation' feature of home-based production, so that cross-border relationships of a spider web type become a dominant feature of modern production technology in many fields.

An important question now is whether all countries and regions will benefit to the same extent from these new opportunities. Will all countries and regions be linked to the spider webs of the modern components and assembly industry? Will new industrial centres emerge or will new technology systems just replicate existing geographical patterns of development? And which are the opportunities of peripheral or isolated regions to acquire their justified share in a 'globalizing' network economy? Is there a case for a 'spider of locational development opportunities' for peripheral areas? And how do new network configurations impact on the competitive position of isolated areas?

This paper will address these questions from the viewpoint of an emerging network economy in our western economies. The main aim of the paper is to investigate the implications of modern network dynamics and of new roles of actors and users in a network for policies concerned with regional isolation and network connectivity. Especially the integration of the European economy will increasingly provoke questions of the nature alluded to above. Therefore, we will in the next section first pay some further attention to networks in a modern society.

## 1.2 Networks in Space

The drive to a network society provokes intriguing questions on the nature and implications of network behaviour and policy. In general, networks may be regarded as "sets of two or more connected exchange relations" (Cook and Emerson 1978). This description incorporates both material (physical) and immaterial networks and refers to the *function* of networks. It might be appropriate to add the *aim* of networks to the above definition, namely that an active interaction is strived for between network users who try to maximize individual benefits from network synergies or externalities (see also Capello 1994).

A glance at recent literature on both industrial organization and transport infrastructure makes it immediately clear that *network connectivity* has become one of the prominent features of industrial economies, as it is able to combine decentralized decision-making with the benefits from synergy. In this context it is often argued that both economies of scale and economies of scope can be exploited to a maximum extent, based on the principle of 'bridging strategies'.

It is evident that there is a variety of different networks, which may be spontaneous, evolutionary or planned. Examples are:

- *physical* networks (in which physical capacity, links, nodes, spatial configuration and service level are dominant features); examples are rail or road.
- *immaterial* networks (in which information plays a critical role through transfer mechanisms such as central facilities, connectivity channels, capacity and receptivity); examples are telecommunication.
- *organizational* networks (in which people act as nodes with many formal and informal linkages and communication channels, organizational objectives and coalition strategies); examples are corporate organizations
- *club* networks (in which network externalities, new information technology and connectivity play a basic role); an example is electronic mail.

All such network types serve to increase benefits of individual users and hence the overall value added generated by network behaviour. Clearly, there may be cases of negative network externalities and exclusion of users, but such cases are less relevant in our context and will not be discussed here. From an equity viewpoint the question can be raised whether and how the transition to a network economy is favourable for peripheral areas. The answer seems to be somewhat ambiguous. On the one hand, competition in a network economy seems to be increasingly concentrated in larger nodes of a network (e.g. mainports, gateways, etc.). In this respect, peripheral areas are in a disadvantaged position which can only be overcome by improving physical infrastructural connectivity or by offering competitive locational advantages (e.g. higher labour productivity). On the other hand, modern forms of network configurations (e.g. telecommunication, CAD/CAM systems, integral logistics) offer many competitive opportunities for network users, so that modern networks do not by definition condemn peripheral regions to an inferior position. Very much depends on the creative self-organization and interaction between actors in such a system.

Clearly, a network is essentially based on *actor dependency*. Kamann and Nijkamp (1991) distinguish in this framework the following dependencies in network relationships: technical dependency, knowledge dependency, continuity dependency, social dependency, logistical and administrative dependency, innovative dependency and financial dependency. It should be emphasized that networks do not have an aim in themselves, but are vehicles for achieving one or more goals of actors using the services or benefits rendered by networks. A necessary condition for the emergence of networks is - as mentioned above - *interaction* between different actors. Since network actors are not usually characterized by spatial juxta-position, network patterns have a clear *geographical* distribution. However, the geographical dimensions of networks are intertwined with order, organization and coordination of such networks. The design, supply and organization of networks should thus meet the *needs of the potential users or customers*. Furthermore, these users have different expectation patterns and needs regarding the services provided by a network, and hence in many cases we observe *different layers* of networks (e.g. roads, railways, waterways, airline connections, telecommunications, etc.). Such networks tend to follow in general the *hierarchical structure* of central and less central places, a situation which may be disadvantageous for peripheral areas in terms of late access to supra-local or supra-regional connectivity. Since there is no single comprehensive network which can serve all needs, we observe in reality a complicated system of partly overlapping, partly complementary network segments. Seen from the user, multiple network configurations may enhance the value added created by networks. Multimodality is a good example of the latter phenomenon, although also here especially the larger multimodal mainports act as central attraction modes to the detriment of smaller places.

The current popularity of network concepts is undeniably connected with the declining domain of public policy: networks tend to become the vehicles through which competition is flourishing. Both external megatrends (e.g. economic integration) and internal system's forces (e.g. new distributional concepts) necessitate a market orientation parallelled by risk minimization strategies. Networks are expected to offer more certainty in terms of expected consequences of strategic decisions (e.g. just-in-time delivery) and hence may be regarded as a major critical success factor in (inter)national competition. The *design and management* of a network should be oriented towards fulfilment of demand conditions of customers (market pull), but it is strongly influenced by technological changes (technology push) and by prevailing property and ownership regimes as well as by geopolitical interests. Thus, competition and regulation are key forces in network development. Since the beginning of this century network supply has often been a public government responsibility, but in recent years - after the recognition of market failures and government failures - we observe an increasingly commercial attitude towards network supply. This means that the organization of the supply side of networks is likely to change drastically with more emphasis on commercial exploitation. This means once more that physically isolated areas (e.g. islands) which have relatively higher costs for their volumes of transport may face more problems in the future.

Beside the socio-economic need for well performing networks, we also witness concern on the *environmental and safety* conditions in physical network planning, construction and use. Network use will most likely have to materialize within ever increasing narrower sustainability limits. The recognition of such barriers requires rigorous social science research on externalities, complementarities and scale economies. The identification and explanation of network performance is in any case a field fraught with many difficulties.

It will not be easy to identify all *success factors of users* of networks, as this depends on a myriad of complex driving forces, such as factor conditions, firm strategies and structures, demand conditions, and relating and supporting industries (see Porter 1990). In a more recent article, Ozawa (1992) distinguishes two types of trade and investment regimes in a network economy, viz. an outward-looking, export-orientated regime focused on external markets and an inward-looking, import-substituting regime focused on protection of home markets. He observes an increasing trend towards the first type of firm behaviour regarding trade and investments, even though in such a deregulated policy environment the government can play an active role in augmenting the market system. This brings us also to the area of network policy.

The set of *network policy* actions that can be envisaged is vast and ranges from direct public supply or intervention to user charge principles or complete laissez-faire. A major challenge of network owners and operators will be to formulate strategic plans that convincingly incorporate non-zero-sum game strategies with gains for all parties involved. This may be illustrated by means of some examples.

The 'user charge' principle in transport policy has in particular become a success in those countries where most suppliers and users of transport infrastructure were enjoying benefits (e.g. suppliers by receiving more revenues from road charges, users by increasing their travel speed, etc.). Likewise the question of intermodal substitution (e.g. from the car or lorry to the train) will critically depend on the willingness to implement such incentives from the viewpoint of win-win strategies for all (or most) actors.

International competitiveness is a necessary condition for enhancing the level of European economic performance after the completion of the internal market. Segmented and nationalistic infrastructure policy may at best serve the shortrun interests of infrastructure owners, but is in the long run to the detriment of all network owners (and users) and affects Europe's economic position. Thus transportation and communication policy requires a balanced implementation of actions which ensure a consideration of both private and social costs, and a global orientation which exceeds country-based or segmented policy strategies. The current plans regarding the European high speed railway system are a clear case of creative action-oriented policy analysis, even though the technology policy underlying this system serves mainly the interest of individual countries.

Networks are at the same time vehicles through which nations (or regions) can control part of the international (or interregional) competition. Monopolistic and oligopolistic structures in space are the result. The socio-economic benefits of coordination and harmonization are often neglected in favour of emphasis on narrow nationalistic interest. This opens much new research in the economic importance of the existence of (deliberate and coincidental) barriers in international networks (including the missing links and missing networks phenomena). This issue, which is particularly important for isolated regions, will be discussed in the next section.

## 1.3 Borders and Barriers: Windows or Impediments?

Industrial restructuring and spatial dynamics are contingent on spatial networks in a mobile society. For many people the 'flying carpet' is a dream which symbolizes a world without any borders, a world with free mobility for everybody, in both central and peripheral areas. Geopolicital and administrative borders are in this perception annoying impediments preventing people from enjoying the fruits of a borderless society.

In our present era we have been witnessing the disappearance of many manmade borders, exemplified by the unexpected opening up and falling apart of the East European power block. At the same time we are observing signs of international cooperation in the EC, EFTA, NAFTA and ASEAN countries which seem to prelude a gradual removal of unnecessary obstacles between the countries concerned, while sometimes installing obstacles with non-member countries. Whether this leads to an open society characterized by the flying carpet still remains to be seen, even though popular terms like globalization, international networking and tele-society seem to suggest that distance does not necessarily mean friction. Also isolated regions have sometimes high expectations on the opportunities offered by the open network society.

Reality however, seems to be fairly harsh, in that vanishing borders do not necessarily imply more openness. Mankind seems to be very keen in inventing new bottlenecks precluding a free movement of people, goods or information. Self interest or group interest is apparently a strong driving force which is often at odds with the social or global benefits of a borderless society and may thus destroy the dream of a flying carpet. The competitive forces in a network economy may even be to the detriment of peripheral regions, a situation recognized in various European countries (e.g. Greece, Ireland and Portugal).

In recent years the recognition has grown that spatial development processes are shaped through a complex interplay of both a heterogeneous physical space and a multi-layer socio-economic stratum. Barriers and borders are evidently playing a very critical but often neglected role in spatial and socio-economic dynamics. Batten and Johansson (1991) offer various interesting historical examples which show that a removal of bottlenecks may have substantial impacts on the growth of regions or nations. For instance, the expansion of trade and the growth of cities in medieval Europe were prompted by drastic improvements in European transport infrastructures (inland waterways, coastal transport, roads) whose dilapidated state had long prevented the emergence of an efficient production and trading system. Similarly, the rise of new commercial centres in Japan in almost the same period would not have taken place without the removal of trade barriers and the creation of new distributional concepts (e.g. scheduled periodic markets, new commercial areas in nodal points of the network, protected castle towns, etc.). The authors suggest that the gradual transition towards a network economy has been of decisive importance for the growth of regions and nations. Especially multi-layer network formation with both interdependencies and hierarchies at all spatial levels seems to become the new model of western economies (see also Camagni 1993). This trend however, evokes at the same time intriguing questions on the position of border areas which are still suffering from distance barriers.

A massive volume of literature has been published on the regional economic impact of infrastructure access. Even though the findings are not always conclusive, it is widely recognized that infrastructure access is a necessary though not sufficient condition for regional growth in a competitive environment. Economic development and infrastructure development go apparently hand in hand (Bruinsma 1994). Therefore, the European economy will remain critically dependent on *well functioning core networks as catalysts for future development*, so that networks become a vehicle for indigenous economic development. There is nowadays however a growing awareness that the current European infrastructure network is becoming outdated, and insufficiently replaced by modern network facilities which would position the European economies at a competitive edge. *Missing networks* emerge because transportation systems are developed in a segmented way, each country seeking for its own solution for each transport mode without keeping an eye on the synergetic effects of a coordinated design and use of advanced infrastructures (Nijkamp et al. 1993). Another reason for missing networks is the focus on hardware and the neglect of software and organizational aspects as well as financial and ecological implications. Market regulations leading to cabotage, protection of national carriers, segmented European railway companies, and lack of multi-modal transport strategies are but a few examples of the existence of low performing European networks. A European orientation of all transport modes is necessary to cope with the current problems of missing and competing networks. It is therefore of great importance that the idea of Trans-European Networks is nowadays strongly advocated by the Commission. But it is equally important that connectivity with respect to isolated areas is maintained as a basic objective in a network society.

*Impediments* to network formation deserve therefore much attention (cf. Batten and Törnqvist 1990). Such impediments may take different forms ranging from institutional inertia or bureaucracy to lack of physical resources or missing infrastructure links. It is increasingly recognized that the rise and fall of nations or regions is often the result of overcoming these impediments (cf. Denison 1967; Olson 1982). This also forms the background of the deregulation movement which seeks to optimize network performance by removing unnecessary bottlenecks of a regulatory nature. Liberalization seems to become the trademark of a network society. Some authors (e.g. Fukuyama 1989) even speak of the 'end of history' as a stage in the history of the western world in which free competition and liberalism are the ultimate driving forces of the modern network economy. Whether this is entirely true for European networks, still remains to be seen, as networks will only operate efficiently if some regulations or agreements (e.g. standardization) are made.

As a result of the deregulation movement, we witness also more emphasis on the *supply and regululatory* side of a network. Hence, the idea of *design* of a network seems to be more relevant than the notion of *control*: design is a structuring activity oriented towards change of form (including barriers), whereas control refers to optimization of key parameters under a given set of *fixed* constraints (or barriers). Peripheral areas are of course more interested in design policies which aim at removing barriers in accessibility. In this context, a barrier may have different meanings (cf. Nijkamp et al. 1990). According to the Oxford Dictionary a *barrier* is an obstacle or circumstance that keeps people or things apart, or prevents communication. The related term of a *border* has more a geo-political meaning: it is the line separating two political or geographical areas, especially countries. It is now an interesting research question what the relationship is between barriers and borders on the one hand and network performance reflected in regional development in the long run on the other. Furthermore, the removal of old borders - a situation we have witnessed regularly in the past years - evokes the question whether all related impediments are eliminated at the same time (Molle 1990).

Borders were traditionally regarded as barriers to economic development: it is not surprising that many lagging regions were found near borders with other countries. Borders were in general creating impediments which hampered the economic gains of trade.

In the past years many old borders have vanished and new maps have emerged. Europe, especially has exhibited a fast dynamic in this respect, and other continents (e.g. NAFTA in North-America) are also gradually following the same trend. This means that the ongoing process of socio-economic integration and economic competition in an open network economy is creating new roles and new possibilities for national states, cities and regions. Barriers related to former borders may disappear, but national self-interest may create new barriers. Thus renewal and establishment are coping with one another.

Governments find themselves in a difficult position, as the deregulations paradigm prevents them from a direct intervention. Controllability via public agencies becomes thus more and more problematic. Cities and regions tend to form their own strategic alliances without too much consideration given to the former borders of nation states. At the same time it has to be recognized that transborder cooperation may generate unexpected benefits, as the economies of scale of new strategic alliances across the borders are significant (see Ratti and Reichman 1993). Consequently, borders are not always barriers to development, but also windows of opportunities. This does not only hold for commercial activities, but also for exchange of information and knowledge (cf. Schott 1988). Thus, such barriers may not only be impermeable and detrimental to development, but also semi-permeable and stimulating for development (e.g. the Swiss watch industry in the Jura) (Ratti and Reichman 1993).

However, in many cases we observe that borders and barriers lead to a lower performance of a network; a border because of geopolitical reasons and a barrier because of institutional, physical or human-made impediments (see Nijkamp 1994). They form an obstacle in a free transfer of people, goods or information. Clearly, some of these impediments are given by nature (e.g. mountains, lakes), but most of them are man-made and created for the sake of convenience or protection or are unintended effects or spinoffs of other barriers. Examples of man-made barriers are: congestion, fiscal constraints, institutional rules, technical conditions, market regulations, cultural inertia, language barriers or information shortage.

It is clear that spatial networks have become the modern vehicles for spatial transfer processes (either material such as persons or commodities, or immaterial such as information knowledge). In this context, Bennett and Chorley (1978) distinguish four types of spatial transfer processes:

- barrier processes: spatial dispersion is limited by obstacles in space
- hierarchy processes: spatial transfer follows a top-down spatial distribution pattern
- network processes: spatial dispersion follows the form of a spatial network
- contiguity processes: spatial dispersion is dependent on contiguity patterns between different areas.

Thus, the morphology of a spatial system creates spatial processes and distribution patterns. The type of spatial development process is contingent upon the shape, the nature and the filters of a spatial network. Isolated areas tend to be more limited in their development potentials than central and well connected areas, especially if both areas are dependent on physical infrastructure. Modern information and telecommunication networks are in principle able to overcome the isolation faced by peripheral areas, but an improvement of their competitive position is not only determined by the available hard ware, but also - and even more important so - by efficient use of software (in a broad sense) and org-ware (especially managerial skills).

The removal of bottlenecks in a network may have profound impacts on the spatial perception of distance and the resultant development consequences. For example, if one compares the European railway map of 1987 - in terms of time distances - with the map in 2015 (after the widescale introduction of high speed trains), one obtains an interesting picture of Europe (see for details Nijkamp 1993) which shows that Europe is getting smaller and smaller, at least for those areas that are well connected.

Because of globalization and other megatrends (including the need for higher and sustained economic growth), transportation in many countries has grown enormously, especially in recent years. This holds for both passenger and freight flows. As the supply of new infrastructure - for various reasons - followed this trend only to a limited extent, existing infrastructure bottlenecks have been accentuated, leading to problem situations almost everywhere. This is a very serious problem, since economic development and infrastructural development have always been strongly interlinked, as they were dependent on one another (see Bruinsma 1994). The full benefits of a network economy will only be reaped in the case of effective (physical and non-physical) infrastructural adjustments. What is needed in a European context, is supranational - and *not* national - thinking and action in infrastructural policy, based on knowledge of past successes and failures in infrastructural planning and of the future needs of the economy, and the constraints imposed by an (increasingly threatened) (natural) environment. This emphasizes once more the need for isolated areas to invest in (tele)communication networks as a way of overcoming isolation.

The role of transport and communication networks for spatial development is indeed rapidly changing. Not many years ago it was possible to argue that - due to underpriced energy and the ubiquity of the road network - accessibility was losing its importance for the organization of space. Today a number of developments work in the opposite direction, i.e. enhance the key function of transport and communications networks for spatial development. Some examples are:

- New, faster transport networks (high-speed rail and air) superimpose new, higher levels of infrastructure on top of the existing hierarchy of transport networks and create new spatial concentrations of accessibility, as is witnessed in the new Trans-European Networks maps of the European Union.
- Strategic megaprojects (such as the Channel Tunnel, the Scandinavian links across the Belt and Sound, the transalpine base tunnels and the conversion of the Iberian mainlines to standard gauge) will create global network connections.
- Despite the reintegration of Eastern Europe into the European transport network, the difference in network accessibility between West and East will for many years remain large. There is still a 'network curtain' in Europe.
- New high-speed, high capacity telecommunication networks such as ISDN and satellite communication complement existing communication networks and create areas of high informational accessibility at nodal centres (see Capello 1994).

The supply-side developments have spawned an unprecedented growth in the demand for international, interregional and intra-regional goods transport, passenger travel and information exchange, have led to the globalization and internationalization of economic processes and have increased the dependency of regions and cities on access to transport and information networks. Borders and barriers are from this perspective a serious threat to a competitive economy, unless these impediments are used as strategic weapons for a new competitive position. For instance, isolated areas tend to become increasingly international tourist resorts.

It does not need much courage nor fantasy to claim that the context and substance of network planning has dramatically changed in the past years.

The *external environment* has a completely different 'face' compared to a decade ago. The drastic political changes in centrally planned countries, the return to market oriented societies and the belief in competition and free entrepreneurship have destroyed the idea that public governments were the vehicles par excellence for ensuring 'the greatest welfare for the greatest number of people'. Regions tend to play a more important role in developing their own future; such a self-reliance holds also for peripheral areas.

Furthermore, the *substance of planning* has changed. The trend to more market orientation in a network society also needs fine tuning to democratic desires of citizens, making network planning a theatre of democratic operations. Thus external megatrends and internal system's movements force planning towards a client orientation. The current trend toward subsidiarity gives may new opportunities to peripheral areas which in the past tended to be dependent on the decision powers elsewhere.

In light of the previous observations, it is clear that there is a broad spectrum of questions which need to be addressed in policy formulation. The recognition of frictions and bottlenecks is the first stage in a policy life cycle. In the past years where many countries have increasingly been faced with the negative externalities of the transport sector, social science research has been of critical importance for formulating issues that needed to be addressed in policy analysis. A simple illustration may clarify this statement. The 'undesirable' outcome of a highly mobile society is - almost paradoxically - the result of rational and plausible actions of a great many individuals. Social science research has convincingly demonstrated that the neglect of social costs in individual decision-making must by necessity lead to a macro outcome that is far from optimal (the so-called social dilemma). This explains worsening quality of life conditions in major cities all over the world. At the same time cities are becoming nodes in a global network (Sassen 1991), especially because of their information gathering, processing and distributing power (Castells 1989). Thus, there is need for peripheral regions to develop pro-active strategies for accelerated development in a modern network society (e.g. on the potential of telecommunication; see Capello 1994). Such pro-active strategies require creative policy and social research, not only regarding technical solutions or financial means, but also regarding material resources, human responses, etc. Those regions which have been able to develop and support such strategies are likely to be successful in their policies. A particularly important, but often neglected factor in this context is the organizational and managerial setting that is necessary for making a policy strategy successful.

## 1.4 Borders and Barriers in Light of the Dynamics of European Networks

It does not need much argumentation to assert that in the past years Europe has been showing the signs of an extremely dynamic socio-economic and geopolitical development: changes in economic heartlands, opening up of Eastern Europe, closer links between EC and EFTA countries, splitting up of former unified countries, strong tendencies towards regional autonomy and drastic changes of former border areas. Traditional centres of economic activity have to compete increasingly with medium-size locations offering economies of scope based on neo-fordist and flexible production systems. Besides, recent developments in new information technology may generate substantial benefits for less favoured regions of the Community, as they do not suffer from diseconomies of scale in the congested European heartlands.

The previous observations imply that the imbalances between different regions in the European continent will not only create a more mobile labour pool (including international and interregional labour migration), but will also create a different - and more promising - scene for the European border regions. In this spatial-economic restructuring process two eminent forces call for more strategic attention, viz. the role of infrastructure and the geo-political changes at the regional level.

As far as *infrastructure* is concerned, it is increasingly recognized that transport developments place a massive burden and an unprecedented tension on a balanced (co-evolutionary and sustainable) development of regions in Europe. The needs of lagging areas, congestion in the European heartland and in almost all metropolitan areas, an ever increasing mobility of people, environmental stress by increasing transport volumes, landscape deterioration by new infrastructure and absence of politically accepted sound economic market principles in the transport sector necessitate policy-makers at all levels to undertake emergency strategies in which integration and coherence benefits are neglected in favour of short-term successes for a limited territory. The discussion on 'missing links' and 'missing networks' is illustrative for this situation, even though the plans for new transnational European infrastructure begin to bear increasingly the fruits of a coherent network. The problem is however, that the relative position of lagged (often peripheral) regions is increasingly worsened by the development of high speed links which favour only the economic centres of Europe. Thus there is a case for the rapid development of tailor-made communication networks for isolated areas in the European space.

Another issue concerns the financing of new infrastructure. The trend towards international network infrastructure in Europe places a high financial burden on many countries. The European Investment Bank can only partly alleviate such financial problems. At local and regional levels private financing has become a more popular vehicle, but also this option does not necessarily guarantee the efficient use of scarce resources (see Nijkamp and Rienstra 1995).

The *geo-political changes at the regional level* do not only concern the position of European centres (e.g. the shift from Bonn to Berlin, or the emergence of new capital cities in the former Yugoslavia and USSR), but also the former border areas. The internal border areas in the EC are likely to receive a sudden improvement in their competitive position in view of their shift from geographical 'dead ends' of a country to new gateways, but the external border areas do not have such perspectives, so that their peripheral position may even be aggravated as a result of more integration and cohesion inside the EC, unless new transnational networks to the East are being built. Furthermore rural areas and coastal areas and islands will also be facing many new challenges with a clear perspective on a structural better position in the 'Europe of regions' (see EC 1991, Amsterdams Historisch Genootschap 1992). If such areas are unable to take their share in the European unification process, the European integration will fail.

In this context, it is also noteworthy that the past years have demonstrated the inability of large organizations to effectively control individuals, groups and regions via a centralized decision system. Again many examples from Europe, Asia and America convincingly show the failure of the nation-state as a model for ruling culturally different people. The recent history of Europe has also demonstrated the potential and viability of the city state or region state as an area with a cultural identity which is more suitable as an efficient decision unit than conventional institutional configurations mirrored in the nation-state. Current geo-political and socio-economic restructuring phenomena tend to eliminate existing borders and to cope with prevailing barriers with the aim to be more competitive in a global economy. Borders and barriers are thus key words in understanding the dynamics in our economies which tend to move towards open international networks. The friction between new borders and old barriers is therefore an important policy and research issue.

It seems indeed to be inevitable that the current geopolitical and socio-economic dynamics will dramatically influence the traditional role of borders between regions or nation-states. In many regions and countries the economic meaning of borders is changing, although this does not imply that a 'borderless' economy is emerging. Changing borders provoke at the same time new issues which are of a socio-cultural and politico-historical nature, such as the sense of social identity, preservation of life style, economic survival, community sense and language. The pathway towards an open network may thus create new barriers which may shape maps of regions and nations which may differ significantly from former administrative delimitations. Peripheral areas tend to have, in general, a well developed sense of cultural identity and this may offer such areas new opportunities for acquiring a more recognized profile in the European setting.

## 1.5 Focal Points of European Network Policy

Europe is composed of diverse regions, both central and peripheral, both overdeveloped and underdeveloped. Therefore, a European network policy should be oriented towards the needs of all regions in Europe. Thus, the future of a unified Europe will be critically dependent on the functioning of strategic infrastructure networks which are *interconnected* in terms of (1) *integration* between different layers of a network (e.g. coordination of high speed/long distance networks such as TGV or airplane and lower speed local networks such as light rail or roads), and (2) *intermodality* between different competing or complementary network modalities in all regions in the European space. In this respect also the quality of *nodal centres* (terminals, stations, urban centres) plays an important role, as well as the frequencies of different types of transport (or carriers) in Europe. This also applies to (tele)communication networks.

The complementary notion of *interoperability* of networks, as advocated in the Maastricht Treaty, generates a new series of important issues which deserve thorough attention from the side of policy-makers and the research community:

- the operation of transnational networks, seen from the viewpoint of European cohesion and East-European (re)integration of all regions
- the close connection between the development of transport networks and (tele)communication networks (including new logistical systems) and their potential implications for the European space (e.g. polarization tendencies towards larger metropolitan areas)
- the interconnectivity of high speed long distance networks and new regional-local infrastructures in central and less central nodes of the European network
- the role of physical barriers (and organizational impediments) in various regions which may reduce the benefits of economic integration in Europe (including the connections with Eastern Europe)
- the emerging conflict between environmental sustainability, infrastructure expansion and competing networks (notably competing transport modes)
- the impact of new transportation, logistics and (tele)communications technologies on network access in the European space
- the lack of standardization of transport systems technologies in Europe, which hamper the full benefits of an interoperable European network, not only for central areas but also for peripheral areas.
- the completely different financing regimes for European transport modes, which prevent fair competition
- the lack of strategic insight into the linkage between European networks and global networks developed in other regions outside Europe.

The previous ideas show that the policy agenda for interoperable European networks is vast and full of scientific challenges. For the time being, there seems to be reason for justified concerns on the future of isolated regions in an open European space. Peripherality means by definition less access to networks and the question is whether such barriers can be overcome only by investing in (tele)communication infrastructure. Even though implementation of the latter strategy is necessary, there is a need for complementary initiatives, e.g. by addressing the strong points of such areas (e.g. as tourist areas, resource areas, etc.). Training programmes and skills are then likely the most effective measures for improving the competitive profile of isolated regions.

## Acknowledgement

The author wishes to thank Yossi Berechman (Tel Aviv) for constructive comments on a first draft of this paper.

## References

Amsterdams Historisch Genootschap (1992) The United States of Europe. Amsterdam

Batten DF, Törnqvist G (1990) Multilevel Network Barriers. The Methodological Challenge. Annals of Regional Science 24:271-287

Batten DF, Johansson B (1991) Origins of the Network Economy. Paper 31st European Congress Regional Science Association, Lisbon, August

Bennett R, Chorley RJ (1978) Environmental Systems. Philosophy, Analysis and Control. Princeton University Press, Princeton

Bruinsma FR (1994) De invloed van Transportinfrastructuur op Ruimtelijke Patronen. Netherlands Geographical Studies, Amsterdam

Camagni RP (1993) From City Hierarchy to City Network, Structure and Change in the Space Economy. Lakshmanan TR, Nijkamp P (Eds) Springer, Berlin 66-90

Capello R (1994) Spatial Economic Analysis of Telecommunications Network Externalities. Avebury, Aldershot, UK

Castells M (1989) The Informational City. Basil Blackwell, Oxford

Cook KS, Emerson RM (1978) Power, Equity and Commitment in Exchange Networks. American Sociological 13:721-739

Denison E (1967) Why Growth Rates Differ. Brookings Institution, Washington D.C.

EC (1991) Europe 2000, DG Regional Policy. Brussels

Fukuyama F (Summer) The End of History? The National Interest 3-18

Hall DR (1991) Impacts of Economic and Political Transition on the Transport Geography of Central and Eastern Europe, Journal of Transport Geography 1:20-35

Kamann DJ, Nijkamp P (1991) Technogenesis. Origin and Diffusion in a Turbulent Environment. Technological Forecasting and Social Change 39:45-46

Lagendijk A (1993) The Internationalisation of the Spanish Automobile Industry. Thesis Publishers, Amsterdam

Molle W (1990) The Economics of European Integration. Dartmouth, Aldershot

Nijkamp P, Rietveld P, Salomon I (1990) Barriers in Spatial Interactions and Communications. A Conceptual Exploration. Annals of Regional Science 24:237-252

Nijkamp P (1993) Towards a Network of Regions. European Planning Studies 1:149-168

Nijkamp P, Vleugel PJ, Maggi R, Masser I (1993) Missing Transport Networks in Europe. Avebury, Aldershot, UK

Nijkamp P, Rienstra S (1995 forthcoming) Private Sector Involvement in Financing and Operating Transport Infrastructure. Annals of Regional Science 29:1995, forthcoming

Olson M (1982) The Rise and Decline of Nations. Yale University Press, New Haven

Ozawa T (1992) Foreign Direct Investment and Economic Development, Transnational Corporations 1:27-54

Porter M (1990) The Competitive Advantage of Nations. Free Press, New York

Ratti R, Reichman S (1993) Spatial Effects of Borders. Europe on the Move Nijkamp P (Ed) Avebury, Aldershot, UK, 115-138

Sassen S (1991) The Global City, New York, London, Tokyo. Princeton University Press, Princeton

Schott T (1988) International Influence in Science. Beyond Center and Periphery. Social Science Research 17:219-238

# 2 Some Notes on Interconnectivity in Transport Networks

P. Rietveld

## 2.1 Introduction

Peripherality and isolation of a region (or a country) have often been analyzed from the viewpoint of physical distance to the geographical centre of a country (or a continent). Much less attention has been given to the qualitative aspects of connectivity between peripheral areas and the entire country. This means essentially that the issue of interconnectivity - in a broad sense the potential access to all relevant nodes in a network - is much more complicated than just the notion of physical distance friction: the quality and structure of an entire network is at stake here, so that one has to investigate the features of both nodes and their connecting links.

An important motivation for the study of interconnectivity is that the quality of transport networks does not only depend on the features of the links, but also on the way the links are connected. An improvement of the connectivity within and between networks may be a much more cost effective way of increasing overall network quality than an improvement of the links themselves.

This notion of interconnectivity is important at several spatial levels, ranging from global, to continental, national and local. In the European context for example it is of relevance given the programmes to construct Trans European Networks. At this level major attention is given to improving the links between networks of different countries, especially where physical barriers exist: mountains and straits. But it is also important when one considers the links between the major infrastructure and the underlying infrastructure. There are often major interdependencies between these, and neglecting them would be harmful for the overall system. A high level type of infrastructure can only function when it is well connected with underlying networks.

It is remarkable, however, that the concept of interconnectivity is often used in a rather vague way. In the present note I address a number of issues to improve our understanding of this notion. For a broader discussion of the topic I refer to Capineri (1994).

The following issues will be dealt with in this paper. First, I will discuss some aspects of unimodal transport systems that are relevant for the analysis of multimodal transport chains (Section 2.2). This is followed by a discussion of the constituent components of the notion of interconnectivity (Section 2.3). Section 2.4 is devoted to aggregation issues, since although interconnectivity is basically defined at the level of an individual node, in many cases one is especially interested in aggregation across for example nodes or transport modes. Quality aspects of interconnectivity are discussed in Section 2.5, whereas special attention to waiting times is given in Section 2.6. Since interconnectivity basically relates to several transport modes, one may observe may different actors in a transport chain. This issue is finally addressed in Section 2.7.

## 2.2 Features of Transport Modes

Transport modes differ largely in terms of cost per passenger (or ton) km and speed (see Figure 2.1). The relative weights attached to both criteria depend stongly on the features of the users (in terms of monetary and time budgets) and on features of the trip itself (distance to be travelled). Thus, depending on the circumstances and on personal features, people will make their modal choice. With unrestricted supply of services for all modes there would be little reason why multimodal trips would be made. In reality, the assumption of unrestricted supply is not valid, however. The level of availability of the services of different transport modes varies largely in terms of space and time (see Figure 2.2). It appears that services which have attractive features in terms of high speeds (planes) or low costs (seaborne freight traffic) are only available at certain discrete points in space and time.

The *discreteness in space* means that multimodal chains are needed to link fine-meshed transport networks with broad-meshed networks at certain nodes in space. The *discreteness in time* means that waiting times will occur at the nodes. These waiting times are among the factors determining the quality of interconnectivity as will be indicated below.

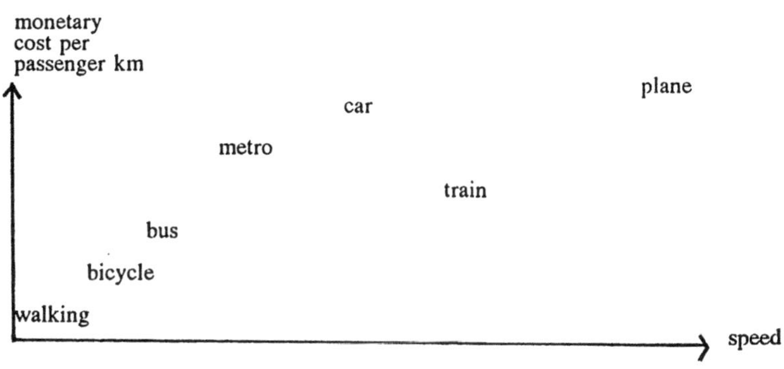


**Fig. 2.1.** Cost and speed features of alternative transport modes

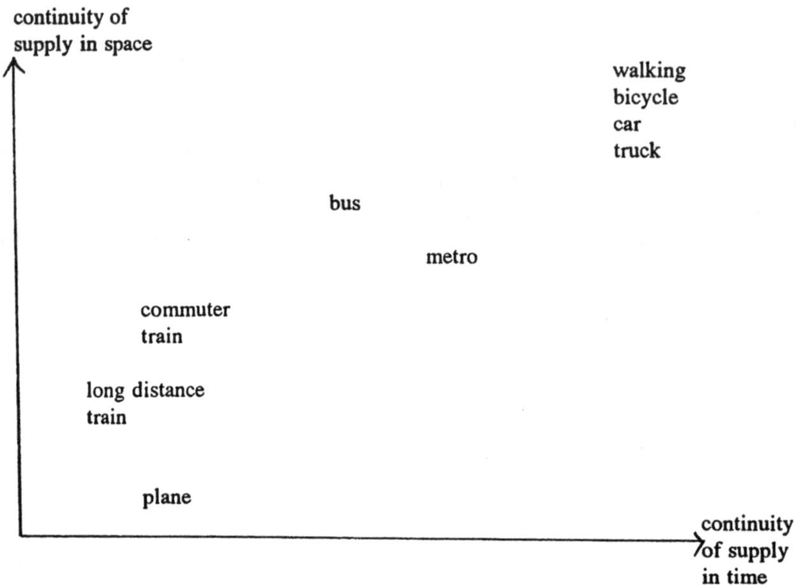


**Fig. 2.2.** Continuity of supply of transport modes in space and time

Confrontation of transport modes according to continuity in time leads to four different cases for waiting times (Table 2.1). Transfer between two modes which are available in continuous time (e.g., car, walking) leads to zero waiting times. When two discrete modes are connected one will have to wait at the node during both directions of a return trip. When only one mode is discrete, waiting takes place in only one direction of a return trip.

**Table 2.1.** Structure of waiting time depending on continuity and discreteness on time

| mode 1 / mode 2 | continuous | discrete |
|---|---|---|
| continuous | no waiting time | waiting time in one direction |
| discrete | waiting time in the other direction | waiting time in both directions |

A similar structure can be formed for the spatial aspects of nodal interchange (see Table 2.2).

**Table 2.2.** Location of nodes depending on continuity and discreteness in space

| mode 1<br>mode 2 | continuous | discrete |
|---|---|---|
| continuous | nodes at any place | location of nodes dictated by discrete space mode |
| discrete | location of nodes dictated by discrete space mode | location of nodes is intersection of both modal networks |

When two continuous nodes are combined (walking and bicycling) one can have nodes at any place. The same holds true for the combination of car and walking, although here, parking policies may give the car system a more discrete character. When two discrete space systems are to be connected (for example train and bus), the location of the nodes is the intersection of both networks.

The above discussion refers to multimodal interconnectivity of networks. However, interconnectivity is also important within unimodal networks. The reason is partly related to the case of multiple carriers. When there are several carriers, users of the network may have to combine the services of several of them. But even when there is only one carrier in a network (e.g. a national railway company) interconnectivity is important. As has been widely discussed in the literature, hub and spoke systems and systems with different service levels (high speed, little stops, versus lower speeds, many stops) give a higher overall quality of services. But this higher quality can only be achieved with an appropriate level of interconnectivity within the system.

## 2.3 The Constituent Elements of Interconnectivity

Interconnectivity is a property of combinations of links and nodes in networks.

To arrive at a more precise definition, I propose to use the following elements, shown in Figure 2.3.

| | | | | | | |
|---|---|---|---|---|---|---|
| | departure time td(x) ↓ | | arrival time ta(y) ↓ | departure time td(y) ↓ | | arrival time ta(z) ↓ |
| origin x | | → | node y | | → | destination z |
| | | mode a | | | mode b | |

**Fig. 2.3.** Constituent elements of interconnectivity

*Interconnectivity* is a property of the following combination of concepts:
- origin x
- link xy : mode a
- node y
- link yz : mode b
- destination z
- at some moment td(x), ta(y), td(y), ta(z)

For example, interconnectivity relates to the possibility to take the train to the city of Frankfurt for the passengers arriving at Frankfurt airport from London by plane at 10.00 p.m. with flight LH 111.

## 2.4 Aggregation of Interconnectivity

In many cases, we are interested in more aggregate concepts of interconnectivity. For example, for European transport policies, one needs information on the degree of interconnectivity between networks in various territories or between a network with international links and the national transport system. This leads to the need of aggregation across one or more of the constituent elements. Some examples are shown in Table 2.3. Case 1 describes unimodal interconnectivity between x and z via y with arrivals in y at ta(y). In case 2 we have the same concept in bimodal form. Case 3 aggregates across a relevant group of origins and arrival times. In case 4 the same is done for the return trip. Case 5 gives an example of aggregation across modes. A high level of aggregation is applied in case 6. Case 7 relates to interconnectivity between a certain origin destination pair (x, z) with given modes via all relevant intermediate points. Case 8 refers to interconnectivity between transport modes a and b in well defined territories of origins and destinations. Of course, many more combinations can be formed on the basis of this table. An important lesson which can be learned from this discussion that when one starts measuring and comparing interconnectivity, one always must define carefully the ways one aggregates certain constituent elements.

## 2.5 Quality and Costs of Interconnectivity: Components

In the above sections we have defined the constituent elements of interconnectivity. Interconnectivity itself has not yet been defined. In its most basic form interconnectivity can be defined as the possibility to travel from x to y with a certain mode a and to arrive at the final destination z with a mode b at a certain moment in time. If such a possibility exists we can say that the pertaining interconnectivity equals 1, otherwise it is zero. This is a rather crude

**Table 2.3.** Possible aggregations of the constituent elements of interconnectivity.

| | origin $x^*$ | inter-mediate point $y^*$ | destina-tion $z^*$ | mode $x^*y^*$ | mode $y^*z^*$ | time |
|---|---|---|---|---|---|---|
| 1. | x | y | z | a | a | ta(y) |
| 2. | x | y | z | a | b | ta(y) |
| 3. | $\bigcup x^* \in X$ | y | z | a | b | $t_1 \leq ta^*(y) \leq t_2$ |
| 4. | x | y | $\bigcup z^* \in Z$ | b | a | $t_1 \leq ta^*(z) \leq t_2$ |
| 5. | x | y | z | $\bigcup a^* \in A$ | b | $t_1 \leq td^*(y) \leq t_2$ |
| 6. | $\bigcup x^* \in X$ | y | $\bigcup z^* \in Z$ | $\bigcup a^* \in A$ | $\bigcup b^* \in A$ | $t_1 \leq ta^*(z) \leq t_2$ |
| 7. | x | $\bigcup y^* \in Y$ | z | a | b | $t_1 \leq td^*(x) \leq t_2$ |
| 8. | $\bigcup x^* \in X$ | $\bigcup y^* \in Y$ | $\bigcup z^* \in Z$ | a | b | $t_1 \leq ta^*(x) \leq t_2$ |

way of dealing with interconnectivity, however, since it does not pay attention to the *cost and quality aspects of interconnectivity*. Important components of the cost and quality of interconnectivity are:

- *Costs of using the intermediate point y.* In container terminals for example, users have to pay for the use of cranes to move the container from one carrier to the other. Another example is provided by airports where airlines have to pay for slots.
- *Costs of storage.* In both airports and seaports, users have to pay for the use of storage capacity of freight depending on the duration of storage. Similarly, one usually has to pay for cars stored at a parking lot depending on the duration.
- *Waiting costs.* Several types of waiting costs can be distinguished. First, waiting relates to the time between arriving at the node y with mode a and leaving it with mode b (td(y)-ta(y)).

  For passengers these costs occur in the form of time losses which can be evaluated in monetary terms by means of value of time estimates. For freight, long waiting times lead to higher stocks and hence higher stock related costs in the form of interest, probability of damage, loss of quality, etc.

  Another type of waiting costs will occur when node y is congested. For example, planes may arrive and leave late at airports because of congestion. Also parking lots may give rise to congestion when they are near their capacity.
- *Inconvenience costs.* The change from one mode to the other may lead to various types of inconveniences which are not yet captured by costs above. These costs may relate to the lack of comfort, seats, agreeable temperature, etc. at public transport stops. Other forms of inconvenience concern lack of information on delays and lack of safety at for example metro stations. There are also compensating features, however, such as the supply of attractive or lowly priced facilities at airports.

- *(Lack of) reliability* is also an important factor determining the quality of interconnectivity at a certain node y. Low reliability should be distinguished from long average waiting times. These have already been discussed above. The problem with low reliability is that this leads to high *variances* in waiting times which may impose additional costs when there is a high penalty associated with arriving late at the final destination.

The above cost and quality components of interconnectivity refer to the users of nodes. In addition to these, one has to take into account costs borne by the suppliers of nodal services such as port authorities, railway companies, firms operating container terminals, public or private firms operating parking lots etc.

## 2.6 Waiting Times and the Quality of Interconnectivity

Time is an important element of generalized transport costs. Therefore, it is important to pay attention to the waiting time aspects of using transport modes.

Consider the case of two airports where one wants to compare the quality of interconnectivity of flights. This can be done by defining a connection between two flights when the interval between arrival of the first and departure of the second is more than a certain critical level (e.g. 45 minutes). Then one can compute the average waiting time for all relevant destination-origin pairs which are served by each airport. Large airports with high frequency connections with many important origins and destinations have the potential to reach low average waiting times. Whether or not this potential is reached depends on the planning of the airlines and the capacity of the airports. The lower level of average waiting time (net of the critical level) is zero. A meaningful upper level would be the average waiting time obtained when the planes would arrive and depart at random moments in time (given their frequencies). The better the coordination of the flights, the further the actual average waiting time will be removed from the average based on the random pattern. As indicated in Figure 4, averages based on random arrivals are lower in large airports compared with smaller airports. However, when users of smaller airports achieve a better coordination of flights their actual average may be lower.

A similar approach can be followed when one wants to study the interconnectivity of two transport systems in a region, for example between the railway and bus system. For each railway station one can compute the average waiting time for people changing from bus to train or vice versa. Aggregation across all railway stations in the region leads to an average value for the region as a whole.

These computations are based on the assumption that transport services are operated according to schedule. The analysis can be strengthened by taking into account unreliability in services. This leads to an increase in the average waiting time and also in its variance.

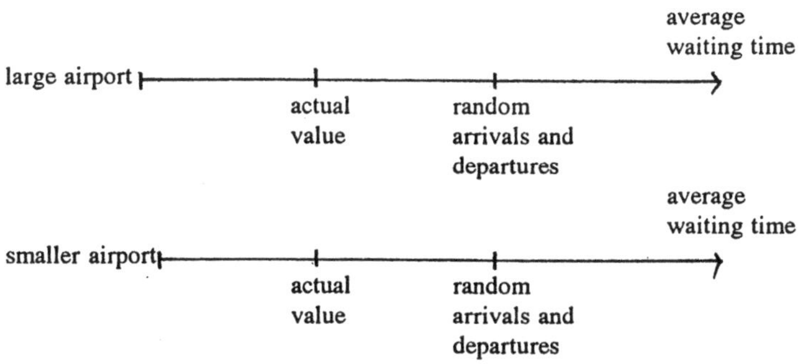


**Fig. 2.4.** Comparing average waiting times of airports

Another relevant aspect of waiting time at nodes in networks is that empirical studies reveal that passengers do not simply take into account the total travel time from origin to destination, but attach different weights to different stages of the trip. For example, based on route choice decisions of Dutch public transport users, Van der Waard (1989) finds that the average public transport user applies a differentiated evaluation scheme. Compared with one minute sitting in a moving public transport vehicle, public transport users apply certain weights to one minute of the following activities:

- one minute walking to a public transport stop is weighed to be equal to 2.2 minutes travelling in a public transport vehicle
- one minute waiting at a public transport stop at the point of entry is weighed as 1.5 minutes travelling in a public transport vehicle
- one minute walking to the final destination is weighed as 1.1 minute travelling in a public transport vehicle
- one minute walking at a change between vehicles (for example from one platform to another) is valued as 2.3 minutes travelling in a public transport vehicle
- one minute waiting at a change between vehicles is valued as 1.3 minutes travelling in a public transport vehicle
- finally, every change between vehicles is valued to be equal to 5.7 minutes travelling in a public transport vehicle.

From these data it is clear that changing from one vehicle to another with a waiting time of 10 minutes will be valued to be equal to about 20 minutes travel time in a public transport vehicle. One implication is that the advantages of the introduction of a rapid service with little stops are not as large as they sometimes seem to be when many passengers have to change vehicle as a consequence of this.

## 2.7 Actors in Interconnected Transport Systems

One of the problems with improving interconnectivity is that there are many actors involved. In general one can distinguish at least seven types of actors:
- the owner of the link between x and y
- the carrier between x and y
- the owner of node y
- the operator of node y
- the owner of the link between y and z
- the carrier between y and z
- the user.

In the case of freight transport the user can be identified as the shipper. With passenger transport the user is usually the person travelling. In Table 2.5 we give some examples of transport chains and the actors in different parts of the chain. Sometimes seven different actors can be observed (A-G). In other cases this is only two or three. In most countries the national railway company owns and operates both the links (and the services on it) and the nodes. In chains where modes are used such as cars, bicycles and walking, the user coincides with the carrier which means that he has a high level of control of the entire system, at least as far as operations are concerned.

**Table 2.5.** Actors in a sample of interconnected systems

| CHAIN | INFRASTRUCTURE OWNER: | | | CARRIER/ OPERATOR | | | USER |
|---|---|---|---|---|---|---|---|
| | xy | y | yz | xy | y | yz | |
| container vessel - container terminal - train | A | B | C | D | B | C | E |
| truck - ferry terminal - ferry | A | B | C | D | B | B | E |
| truck-city distribution centre-small truck | A | B | A | C | B | D | E |
| train - railway station - train | A | A | A | A | A | A | B |
| plane - airport - taxi | A | B | C | D | E | F | G |
| bus - bus station - walking | A | A | A | B | B | C | C |
| car - park and ride - train | A | B | B | C | B | B | C |
| car - parking lot - walking | A | B | A | C | D | C | C |

## 2.8 Concluding Remarks

Interconnectivity is an issue that deserves systematic attention in the analysis of transport networks. Improvement of interconnectivity may be a cost-effective way of increasing the over-all efficiency of multimodal transport networks. For an improvement of interconnectivity technology may be helpful in two ways. The first way relates to the improvement of the technology of transfers in nodes so that transfers can be done in a quicker, cheaper or more convenient way. The second way concerns the use of telematics to improve the quality of nodes, leading to flexible transfer procedures implying lower waiting times.

But technology is not the only means to improve interconnectivity. From the discussion above it is clear that in many cases interconnectivity depends on the behaviour of many actors. Lack of coordination between key actors involved is an important factor leading to a low quality of interconnectivity. Institutional changes are sometimes needed to overcome these problems. Sometimes, these changes may imply the introduction of a larger degree of competition in the markets concerned. It is observed in many countries with highly regulated public transport that suppliers of public transport services do a bad job in terms of coordination of the networks. A stronger degree of competition, for example the possibility of free entry of new suppliers may make incumbent suppliers more sensitive to the wish of passengers to have short time losses while changing.

Lack of free competition due to regulation in transport markets is not the only possible reason of bad interconnectivity in transport networks, however. There are also examples of the opposite case: free competition in transport markets may lead to suboptimal schedules from the viewpoint of users (Rietveld and Rouwendal, 1994).

The analysis of interconnectivity may shed new light on the degree of isolation of peripheral areas and also suggest new ways of policy-making concerning the alleviation of peripherality. Such policies should not only refer to the hardware of connections, but also to regulatory regimes, organizational arrangements, intermodality strategies and technological advances for network infrastructures.

These examples are sufficient to illustrate that the concept of interconnectivity is not only relevant from the viewpoint of transport policies, but also quite interesting from a scientific point of view.

## Acknowledgement

The author thanks Peter Nijkamp for constructive comments and suggestions.

## References

Capineri C (1994) Interconnectivity and network development, some territorial references. University of Siena, Siena

Rietveld P, Rouwendal J (1994) Temporal competition in transport. An equilibrium approach to departure times in public transport. (Mimeo) Vrije Universiteit, Amsterdam

Van der Waard J (1989) Onderzoek weging tijdelementen. (Mimeo) Technical University, Delft

# 3 Accessibility and Peripheral Regions

R. Vickerman

## 3.1 Concepts of Accessibility

The problem of peripherality is essentially one of accessibility. However, accessibility is not a simple concept, changes in accessibility can be made in many different ways and this needs to be understood if changes in the accessibility of a peripheral region are to have any impact on its economy and development potential. Furthermore, simple improvement of accessibility, however defined, gives no guarantee of improved performance. It can open up the local economy to increased competition. At least it is likely to require other policy initiatives to capitalize on any improvements in accessibility.

The message of this paper is that accessibility needs a more careful definition, reflecting a number of cross-cutting dimensions, these measures need to be related to the specific needs of the regional economy, and that this implies caution should be taken with traditional aggregate measures.

The term accessibility, as frequently used, encompasses (at least) two concepts, that of location and that of market potential. Both of these convey useful information and any attempt to assess accessibility needs to include both. It is simpler to start with potential measures and work back to those reflecting location.

### 3.1.1 Concentration of Population

At its simplest, market or economic potential is about access to population. A simple map of Europe showing urban centres or population density shows clearly how population is concentrated in core regions at the geographical heart of the EC, in the area between London, Paris, Frankfurt and Amsterdam. Outside this core lie a number of substantial, largely free-standing, metropolitan regions from Glasgow, Copenhagen and Hamburg in the north to Munich, Milan and Madrid in the south. Each of these provides a substantial concentration of population. Beyond this ring lie much more sparsely populated areas with metropolitan centres such as Dublin, Lisbon and Athens, which are less significant in a European context, but very significant in a national context.

This gives a clear impression of accessibility being related to economic power and performance, but such a conclusion depends on assuming that both distance and population are unambiguous determinants of access costs and spending power. It is these points which we need to take up in more detail.

### 3.1.2 Regional Incomes

The map of regional per capita incomes displays a similar concentration of wealth in core regions of the Community. It is not just a question of large markets, but of wealthy markets. However, the picture is not as clear cut as the distribution of population. There is no unambiguous centre of per capita wealth in the EC with a neat monotonic trend downwards in per capita incomes as we move outwards from the centre. Two particular features stand out here. One is the existence of substantial wealthy concentrations at some distance from the traditional core of the Community. Regions in Southern Germany and Northern Italy, and individual metropolitan areas such as Lyon or Copenhagen stand out. We can also see the way in which this trend may extend to other areas; Barcelona and (if we consider the expansion of the Community to include new members) Vienna would be good examples. The second is the simultaneous existence of relatively poor regions close to the economic heart of the Community. There is a band of regions with per capita incomes below the EC average which runs from the east of London, along the Franco-Belgian border to Lorraine and then along the Belgian and Dutch border with Germany. This includes regions such as Wallonia, within which the Province of Hainaut, just next to Brussels, has a per capita income low enough for it to have been included recently in a new tranche of regions to be granted Objective 1 status. Although these regions are likely to have suffered in the past from their border status, it seems unlikely that this alone can have resulted in per capita incomes up to 25 per cent below the Community average and up to 50 percentage points below the adjacent metropolitan regions.

Proximity to wealth does not seem therefore to be an automatic guarantee of wealth. One of the problems of regions close to major metropolitan areas is that they tend to live in the shadow of the major city. In some cases their lack of wealth may be just a statistical artefact. Commuter regions will, for example, show low levels of income on an output measure (that usually used) since much of the resident population works in the metropolitan region. But this is not a complete explanation. Clearly in other regions, such as Wallonia, industrial structure has a part to play.

At the other extreme, the poverty of the very peripheral regions, such as those of the Aegean, may reflect their remoteness from centres of wealth, but also reflects the lack of development in the nearby metropolitan regions. Thus the primary problem for such a region is in its dependence on Athens and the development problems of the Athenian economy rather than its remoteness from Brussels or Frankfurt. Inter-regional interdependence on the macro-regional scale may thus be very significant as a determinant of a region's fortunes.

### 3.1.3 Time Accessibility

We have dealt with the question of regional incomes which is a measure of market power; the other dimension of this is the measurement of physical accessibility. Potential measures tend to use road distances, adjusted for quality and breaks in the network. Of course the economic justification for this is that we need a measure of cost of access to markets. Increasingly, distance is becoming a less reliable measure of this because of changes in both the quality of transport networks (and the relevance of more than one mode) and because of changes in the nature of the goods being carried. For this reason it can be argued that time is more relevant. Time is at least mode neutral, it can be used as a single measure where more than one mode is used, penalties can easily be directly introduced to allow for breaks in networks, changes of mode and it is possible to reflect quality of service in terms of frequency.

If we consider time accessibility we can see that it may change the pattern of access from that implicit in a distance based measure. More remote regions, especially islands, may benefit because they use air as a normal means of transport. Island regions thus perform better on such a measure than some more remote mainland regions. Thus it is possible, even with connecting flights, to travel from one side of Europe to the other in five to six hours. Many fairly central regions, reliant on land transport due to there being no local airport, may need the same time to cross the core region, either by road or rail or by the time to access the nearest airport to use air. However, we would not conclude that the island regions have as good accessibility, they rely on one mode of transport and face infrequent connections to a limited range of destinations. More importantly they will be perceived as having poor accessibility. Perception may be the most important factor here, governing the views of both inward investors and potential purchasers worried about reliability of supplies.

### 3.1.4 Economic Potential

In the previous two sections we have been exploring the two elements which make up conventional measures of economic potential. In the tradition of Clark (1969) and Keeble (1982, 1988), economic potential is a measure of market area. This is based on the summation, for each region, of the values of all regions' incomes, adjusted by a measure of intervening distance:

$$P_i = \sum_j \frac{y_j}{d_{ij}^2}$$

where $y_j$ is the income of region j and $d_{ij}$ is a measure of the intervening distance from i to j.

Economic potential is designed to give a first impression of market size for a region. Although the authors of such studies give clear caveats about the way their results should be interpreted, the temptation has nevertheless been to use them to produce simple maps which show contours of potential. Naturally economic potential is greatest at a point close to the geographical centre of the EC, since this is close to the highest levels of concentration of income and wealth achieved. The real problem with potential is that it ascribes income levels to regional centroids and uses a distance based deterrence function which assumes a degree of continuity in transport costs. This produces a deceptively uniform reduction in potential as we move away from the centre. It is true that the potential measures mirror fairly well the achieved levels of GDP per capita in peripheral regions (typically less than 20 per cent of those achieved in central regions), but they do not capture the variety of regional experiences in the intervening space. This is confirmed by examining the suggested regional typology which emerges from potential analysis. This too tends to display a degree of concentricity in regional types.

Perhaps the greatest problem in using potential analysis is that it encourages belief in continuing cumulative causation. Why should any firm invest in a region with low potential, especially if it recognises that failure of other firms so to invest would reduce that potential even further? Yet we find that investment does take place in peripheral regions. Part of this is due to the great differences in market area needed by firms in different industries. The achievement of scale economies is critical in determining appropriate market area size. Secondly, the appropriateness of transport provision is important, the combination of unit cost per unit distance, need for speed, reliability, etc. For example, coastal locations in peripheral regions can obtain slow, but reliable, single mode transit to major ports, which may make them an attractive proposition for some products whereas inland regions with similar degrees of peripherality have no such option.

### 3.1.5 Critique of Aggregate Analysis

The above discussion of potential demonstrates that the real problem lies in the degree of aggregation. In order to produce a measure which more accurately reflects the precise impact of a region's location on its market potential we need it to be sensitive to variations in output structure and modal variety in transport provision (Vickerman 1994a).

The output structure of the region is a key determinant of its transport needs in terms of both the export of finished products and the import of components, but also in terms of its internal functioning. In many cases the ability of the region to satisfy its own internal transport needs may be the crucial determinant of the competitiveness of its industries. In core regions, congestion, and in peripheral regions, inadequate basic infrastructure, may lower the efficiency of the transport system making it difficult to ensure deliveries of locally produced supplies and labour and thus raise costs.

It is, however, more complicated than this since similar industries may use substantially different production processes in different locations. Responding to the local supply price of inputs, firms adjust their production technology. Transport is a key input and hence where transport costs are high firms will be encouraged to economize on transport as an input. This leads to the evidence that in some relatively peripheral regions the average level of transport costs is no greater than in more central regions (Chisholm 1987, 1992).

As well as the need for disaggregation by sector, the variety of transport on offer also requires a degree of disaggregation. The lack of a regional airport or seaport naturally precludes easy use of these modes without considering in detail the nature of access to them. Similarly the quality of provision of road and rail in terms of existence of high-speed rail or motorway alternatives will determine the nature of usage of these modes. Furthermore, with sea and air modes, variations in the quality of service will be critical. Quality here is measured in terms of frequency, and the provision of direct services to major destinations. With all modes, there is a need to consider the developments of the market. Privatization, or increasingly commercial objectives being set to state-owned companies, changes the nature of accessibility. Where reliance is placed on relatively few modes this may result in an increasing peripheralisation if alternative sources of subsidy are not introduced. Reductions in scheduled frequencies to obtain higher load factors for commercial reasons may lead to problems in obtaining space on services at short notice. This may be perceived as a substantial loss of accessibility by potential users.

## 3.2 Developments in Transport

We have been dealing so far with the static measure of accessibility. We now need to consider how these static measures respond to changes in transport supply which lead to a change in patterns of transport demand, and potentially to changes in sectoral structure (Vickerman 1994b). We consider three main changes here relating to the growth of high-speed rail, to inter-modal transport interfaces and to changing assumptions of competition between modes.

### 3.2.1 High-Speed Rail

Perhaps one of the most important changes in the transport map of Europe for the 21st century is that provided by the development of high-speed rail networks. During the 1980s we have seen a number of separate routes developed for high speed, based essentially on city pairs. The major changes which will start occurring in the 1990s is the joining of these separate routes into a major network. The links established initially were largely outside the core regions, linking such cities as Lyon, Bordeaux, Hamburg and Munich to the core regions. The new developments will link more such free standing cities into the network, but the major change will be the completion of a substantial new network at the heart of the core region linking London, Paris, Brussels, Amsterdam, Cologne and

Frankfurt. This will have major impacts on both these metropolitan regions and the regions in between, crossed by the new lines, but often without good access.

The questions this poses are whether such a network will increase the process of centralization and whether the response of the peripheral regions should be to demand their own links to such a network. To some extent the high-speed network is a product of the process of centralization. The increasing wealth of the core metropolitan regions has created both the conditions in which a high-speed network can be afforded and the need for it to continue to support the level of economic interchange. However, it is also necessary to avoid this success stifling itself in increased congestion. For the peripheral regions the key may be not to look to high-speed rail as a solution. It is certainly likely to be less relevant to situations where levels of flow are much lower. Even where there is a case for extending the high-speed rail network into more peripheral regions, the critical determinant will be access to the network. It is notable that within France, the key to the accepted success of the TGV network has been its ability to serve a wide range of destinations not on the new lines.

### 3.2.2 Modal Interface

As well as the question of linking inter- and intra-regional networks within a mode, the interfaces between modes are becoming a major issue in determining accessibility. Two examples serve to illustrate this. At the regional level there are the links established to serve major inter-regional links. Urban rail systems serve inter-regional links, as does the need to ensure an adequate park and ride system to enable car users to transfer to other modes. At the inter-regional level there is an increasing interest in linking high-speed rail to airlines. This has the advantage of enabling passengers to transfer between national and international or inter-continental transport systems more easily, thus improving the accessibility of regions not benefitting from the existence of an international airport. Furthermore, by liberating scarce air space and runway capacity by reducing the number of shorter distance movements (up to about 500 km), the efficiency of the air transport system as a whole can be improved.

### 3.2.3 Modal Competition

One of the principle arguments against the essentially planned view of the transport system as a whole, outlined in the previous sections, is that it fails to allow for the benefits of improved competition. There is an increasing interest in seeing transport services as being more self-supporting and, if possible, transferred to the private sector. This is easier to achieve in some cases than in others. For air and sea transport, as with buses, it is easy to split the infrastructure from the operator and to have competing operators using the same infrastructure. For rail this is more difficult to achieve. Thus regions dependent on air and sea, such as island regions (or island nations such as Ireland) will face a potentially different economic regime from those where rail transport dominates. This raises questions of regulation and subsidy. It also focuses on the issue of how far service levels and fare policy should feature in measures of accessibility.

## 3.3 New Measures of Accessibility

In this section we briefly review some of the newer approaches to accessibility measurement and assess possible future developments to incorporate issues raised above.

### 3.3.1 ICON

The ICON index is based on an aggregate measure of the connectivity of a region to the transport networks of the Community (CEDRE 1993). The networks considered include motorways, other high standard roads, railways, commercial ports and airports.

The index uses small regions, of around 500 km$^2$, defined at NUTS level 3. This is important in order to obtain a measure which is representative of a region. With larger regions there are likely to be substantial variations in accessibility between urban and rural areas and between urban areas of different sizes. Even at NUTS 3 there are likely to be unrepresentative measures, and here the measure is based on that for the principal town of each region.

The index concentrates on access in terms of time to the various networks, rather than access along those networks to other regions. It is more concerned with the internal accessibility than the external accessibility of these regions. The connection time to each network is weighted by the relative importance of that network and a measure of the level of service. Ideally the weights should reflect the likely usage of each network by the activities of a region, e.g. for a region where the export of bulk raw materials is important, access to a port is more critical, for a region with a large financial services sector, access to airports and high-speed rail is more important. Penalties are introduced for network discontinuities such as the need to use ferries (e.g. 5 hr for UK; 10 hr for Ireland) or because of rail track gauge differences (3 hr for Portugal and Spain).

The index is used to define levels of connectivity in terms of hours. As would be expected the worst accessibility is found most distant from the major transport corridors and the main centres. In the context of the Atlantic Arc this produces the interesting result that in Britain and Ireland it is the more remote, physically distant, coastal regions which have the poorest connectivity. In Portugal and parts of Spain and France, it is typically interior regions which are worst off. This is due to physical factors such as mountains, but also reflects the trading situation of the economy, especially in the case of Portugal.

Perhaps the major problem with ICON is its definition of accessibility independently of the transport needs of specific regions. The index is quite difficult to interpret. It is not clear whether all modes are used for all regions, e.g. are all regions given a port accessibility index when the absence of a port may be irrelevant to the needs of a region? The cartographic representation of changes in the index does not allow for the importance of the change for populations. Since it is area based it fails to highlight that a major gain in accessibility for a small, but densely populated, urban area should be much more important.

### 3.3.2 Bundesforschungsanstalt für Landeskunde und Raumordnung

The Index developed by the Bundesforschungsanstalt für Landeskunde und Raumordnung (BfLR 1992) is a more focused and a rather more complex index. Its primary function is to provide a measure of accessibility from NUTS level 3 regions to sets of major EC destinations. In total 194 primary centres are chosen, but separate indices are calculated for access (termed the situation index) to this complete set, to the nearest three centres (a measure of the wider regional situation), to the nearest of 34 major centres (a measure of proximity to major agglomerations) and also for accessibility to certain key sites such as nearest high-speed rail station or airport. These indices do not reflect differences in the importance of the various destinations, but accessibility to population is considered in a separate index which measures the population resident within certain time bands (3 hours being taken as the critical cut-off). Population here appears to be ascribed to zonal centroids thus there are some problems with larger regions. This index also encompasses the importance of accessibility to, and accessibility along a network, but this time to defined destinations. Again accessibility is measured in terms of time, allowing for variations in quality along the network and including such elements as airport check-in and transfer times where appropriate. Variations in quality of service are allowed for between countries, e.g. car traffic speeds vary for motorways and some other roads.

A particularly important innovation is the allowance for joint use of modes. Most accessibility indices assume use of a single mode, but efficient transport from more peripheral regions can often be obtained by combining modes. This is difficult to model, especially where it is intended to forecast changes in accessibility resulting from say new infrastructures, but without knowing how individuals will respond to such changes.

The set of Indexes produced by this method provide an interesting picture of the diversity of accessibility. In particular it highlights the relatively good accessibility existing in most metropolitan regions of the Community, regardless of their location. However, there are some interesting dimensions to this; the most peripheral major centres, such as Athens, Lisbon and Dublin are hindered by their remoteness from the next nearest major centre. Accessibility clearly runs along major axes of communication and links to airports and is hindered by physical barriers (especially mountains) and borders.

The Index is based almost entirely on measures relating to business passenger traffic flows. There is some discussion of seaport access, but this is not fully presented. This again begs the question of what is the critical accessibility factor for each individual region. Thus, for example, we again encounter the problem of circularity - regions with good accessibility to metropolitan areas have developed because they have good accessibility and have developed on the basis of sectors which need good accessibility. Improving access to airports or high-speed rail might improve their index, but not change the fundamental situation with respect to their current sectoral structure.

Although the BfLR index is constructed more directly in terms of access to other regions, the same criticism applies as to ICON, in that it is too aggregate and fails to capture the specific transport needs of individual regions. However, the temptation to reduce everything to a single index has been resisted; it is particularly useful to separate out what is termed the 'situation' index (that relating to access to destinations) from the 'accessibility' index (that relating to access to population).

### 3.3.3 A Comparison and Assessment of Indicators

To a large extent the two indices discussed above are complementary rather than competitive. ICON is principally concerned with measuring access to networks, adjusted by the quality of the network at that location. The BfLR Index is concerned with simple questions of access along networks (or more interestingly taking account of alternative modes) to principal destinations or populations. Both use time as the key unit, which has many advantages for dealing with passenger traffic. Neither deals adequately with the question of goods transport, for which time is often less important than reliability, and which has much more specific requirements for different traffic types.

It may be useful to consider ways in which the two indices can be linked or combined when looking at the situation in any region. However, in terms of broad orders of magnitude there is not a great inconsistency between these two indices in giving an impression of the accessibility situation for regions in the Atlantic Arc. Both represent a substantial improvement on the even more aggregate Keeble et al (1988) measures. These bring a much needed shading to the rather crude views of centre and periphery used hitherto. In particular the move from NUTS 2 to NUTS 3 enables a more precise view to be taken. the use of more than one mode is a great improvement - though more needs to be done on investigating appropriate ways of aggregating modes in a way which is relevant to the specific needs of individual regions. Much more needs to be done on measures for freight transport which is more complex in terms of accessibility objectives than is passenger transport. Such developments will be even more data hungry than these measures, but this is important to enable assessment of the impacts of accessibility changes on regions' economic development.

### 3.3.4 Future Developments

Three issues are worthy of specific mention in terms of scope for development:

- *Inter- and intra-regional accessibility*: the need to combine the internal quality of the transport system with the external for a region in order to measure the full extent of accessibility to the region. There is a danger that we concentrate on the accessibility of major gateways when the real constraint on development is internal access to those gateways.

- *Role of freight transport*: since accessibility is most easily measured in terms of time and this is most easily available for passenger travel, freight accessibility gets overlooked in many cases. Freight raises particularly interesting difficulties since multi-modal usage is more common, and although freight shippers are often considered to be excessively committed to a single mode, they are nevertheless more likely to be well-informed and influenced by changes in both accessibility and cost than many passengers.
- *Reliability*: relates both to network penalties and inventories. The issue of network penalties has been discussed in detail above. These are, however, easier to measure than to evaluate. Where transhipments can be built into a schedule they may not impose excessive costs. Where, for example, ferry crossings interfere with through road haulage, they can be taken as rest breaks required by drivers, such that the time penalty is much less than may be apparent. Often the degree of penalty will not depend on the network but on the journey being taken. For example the time penalty imposed by the English Channel is much greater for short journeys between London and Paris or Brussels than for long journeys between Manchester and Milan. Where this impinges on firms is not typically through the level of transport costs, which tend to be relatively low in relation to total costs, but through the effect which variations in these costs (especially unpredictable variations) have on the need for higher levels of inventory holdings than would otherwise be necessary (especially where just-in-time production procedures are in use).

## 3.4 Transport Costs and Other Input Costs

A key issue is the extent to which changes in accessibility, through changes in transport costs give rise to changes in a region's competitiveness. As outlined above, changes due simply to changes in transport costs are likely to be relatively small. Those relating to the way firms use transport and the impact of this on inventory holdings may be of much greater significance. There is a further consideration, however. Firms (and indeed individuals) do not just use transport as the conventional derived demand from a pre-determined level of activity. Variations in the availability and cost of transport can lead to substitution between transport and other inputs, typically resulting in changes in location. For firms in peripheral locations this involves a trade off between higher costs of transport and lower costs of other inputs such as land and labour. Changes in accessibility for peripheral regions may have much more profound and far reaching effects on location decisions than the simple change in costs would indicate.

## 3.5 Policy Conclusions

We have concentrated here on outlining the issues and problems in an adequate definition of accessibility which can explain the likely impact on a region of changes in the provision of transport. In this final section we draw together some of the implications for policy which emerge from this. These can be grouped into four main areas of concern:

- *link with industrial location*: it is clear that transport policy, especially that towards the provision of new infrastructure and services, cannot ignore the wider impacts on industrial location and activity. Transport cannot just be measured in terms of the costs of transport provision to a region's existing industrial structure. From a policy point of view therefore there is a two-way link between transport and regional industry policy.

- *reliance on intermediate regions and countries for network development*: it is clear that networks rather than individual links are the critical factor and hence developments in the network well away from a region may be more significant than those within the region or directly linked to it.

- *fixed and mobile capital*: in terms of developing accessibility both the development of fixed infrastructure and the level of service provided on that infrastructure will be important. For peripheral regions in particular the mobile capital of planes and ships may be of much greater significance than the fixed infrastructure of railways and roads which will dominate the situation in more central regions. From a policy perspective, therefore, it is essential to ensure an equal treatment of the funding arrangements for both fixed and mobile capital.

- *infrastructure and competitiveness in service provision*: from the point of view of the user of transport it is the full cost of the service provided which is significant, including service levels, reliability etc. Whilst the provision of additional infrastructure, by increasing capacity, will have an effect on this cost, the simple provision of additional capacity does not guarantee a reduction in costs. Monopoly control of infrastructure, including inefficiencies resulting from a regulated environment may reduce the potential benefits. Private provision of new infrastructure may create large debt burdens which have to be borne by users, thus negating the potential benefits of the new infrastructure.

Thus there remain many complexities in the evaluation of the impact of new infrastructure provision, especially for the more peripheral regions. The first stage in this evaluation process is further refinement of the definition of indicators to measure all dimensions of accessibility.

## References

Bundesforschungsanstalt für Landeskunde und Raumordnung (1992) Accessibility and Peripherality of Community Regions. The Role of Road, Long Distance Railway and Airport Networks. Report to European Commission

CEDRE (1993) Etude Prospective des Regions Atlantiques. Report to European Commission, Centre Européen du Développement Régional, Strasbourg

Chisholm M (1986) The impact of the Channel Tunnel on the regions of Britain and Europe. The Geographical Journal 152:314-334

Chisholm M (1992) Britain, the European Community and the centralisation of production. Theory and evidence, freight movements. Environment and Planning 24:551-570

Clark C, Wilson F, Bradley J (1969) Industrial location and economic potential in Western Europe. Regional Studies 3:197-212

Keeble D, Owens PL, Thompson C (1982a) Regional accessibility and economic potential in the European Community. Regional Studies 16:419-432

Keeble D, Offord J, Walker S (1988) Peripheral Regions in a Community of Twelve Member States. Office for Official Publications of the European Community, Luxembourg

Vickerman RW (1994a) Regional science and new transport infrastructure. In: Cuadrado Rouro J, Nijkamp P, Salva P (Eds) Moving frontiers. Economic Restructuring, Regional Development and Emerging Networks. Avebury, Aldershot

Vickerman RW (1994b) Transport infrastructure and region building in the European Community. Journal of Common Market Studies 32:1-24

# 4 Innovative Growth and Peripherality in the New European Territory

G. Kafkalas

## 4.1 European Integration and the Reproduction of Peripherality

The process of European integration after the creation of the Single Market and the ratification of the Maastricht Treaty for the political and economic Union of the twelve by 1999, remains a source of extreme geopolitical confusion. The plans for enlargement include as a first step the accession of the formerly EFTA countries, Austria, Norway, Sweden and Finland, while the formerly planned economies such as the new Czech and Slovak republics, Poland and Hungary, are also serious candidates for accession to membership within foreseeable time horizons. Furthermore, the Baltic countries (Latvia, Estonia, Lithuania), the Balkan countries (Albania, Rumania, Bulgaria and the former Yugoslav republics) and the Mediterranean countries (Malta, Cyprus, Turkey) are of major importance to the Community. This situation is reflected in the granting of special agreements and the provision of assistance under various schemes, though there is no common policy for the status of all the above countries in relation to the European Union. All these developments have greatly changed the debate and the prospects concerning the pace of the unification process, the expansion of geographical boundaries and the institutional deepening of the European Union (Holmes 1992; Mitsos 1993).

The further deepening of the twelve is a project favoured by many forces within the present structure of the European Community. This option is argued for on the basis of the assumption that the creation of a single market for capital and commodities without a federal state structure endowed with a comprehensive set of public policies (fiscal, regional, social, etc.) will work against the major objectives of convergence and cohesion and will trigger defensive and protectionist strategies together with an intensification of competition between the countries and the regions of Europe that may threaten the process of unification itself. However, the prospect of a federal Europe has been weakened by the fact that the twelve members strive to combine their different national socio-economic trajectories with the straitjacket of the list of the common austerity requirements introduced by the Maastricht Treaty (Ioakimidis 1993, 466; Andrikopoulou, forthcoming).

As a response to the above, many alternative visions of Europe's future have been created, including some seriously debated models of a two-tier, a homocentric or even a Community of changing geometry, a fortress Europe or an ever-expanding pan-European market. These futures reflect combinations of external and internal divides and alliances which are themselves unstable and are depending upon other combinations of developments occurring within all the prospective partners (Dunford and Kafkalas 1992). The realization of any of the above scenarios is extremely important for all the countries and regions of Europe though the distribution of costs and benefits is not visible in advance. However, the historical outcome acquires critical significance for the less favoured and especially the isolated regions of the european periphery which need large amounts of assistance in order to converge towards the european average and maintain the cohesion of the Community.

The situation is further complicated by the apparently antithetical processes of globalization and localization (O'Brien 1992; Amin and Thrift 1993). Globalization is the powerful process towards a global market of commodities, finance and information strengthened also by the recent GATT agreement. And it is exactly the collective wish of the twelve members of the European Community, in response to the global forces at work and faced with the weakening of the national state, to coordinate their actions in order to increase their bargaining political and economic power vis-a-vis the other global players, notably the US and Japan. This need also guides the pace and the orientation of the various designs towards European unification.

Localization, on the other hand, is a process implying the increasing transfer of powers and initiatives to the subnational, regional and local levels of administration and representation. Localization seems also both to be triggered by and contribute to the erosion of the national state. Both processes, globalization and localization, are thus mutually reinforced components of the transformation of the models of development, sometimes associated with the collapse of Fordism and the emergence of flexible and networked types of development (Storper and Scott 1993). The emerging new models are characterized by the introduction of a wide range of innovative elements which are the decisive factors for the understanding of new growth poles and peripherality in Europe.

There are strong fears that whatever option will be realized for Europe in the course of history, the cumulative and long term consequences of the process of European unification will result in favour of the more developed and against the poor regions of Europe. These fears provide the rationale for cautious movements that combine whatever steps are being taken towards either enlargement and/or deepening with the implementation of counter policies which aim to promote social cohesion and economic convergence within the European Union (Maraveyias 1993). In all cases the question of convergence and cohesion is to be understood in both political and economic terms. The political argument favouring the support of the less favoured regions is that the vision of a European Union, in addition to its ideological appeal, should deliver positive results to all and could not be achieved at the expense of some of its citizens. The economic arguments relate to the exploitation of unused resources and the functionality of the periphery for the development of the european core (Begg 1993).

The reservations concerning the ability of LFRs to adapt in the Single European Market as well as the need to support their assimilation through structural policies are of particular importance for the least endowed among them and in particular the remote and isolated regions in the mountains, the external frontiers of the Community and the islands. It is convincingly argued elsewhere that the benefits from the internal market are not automatically distributed throughout the european territory (Dunford 1992), while the existing regional inequalities pose serious problems for the assimilation of LFRs and especially the remote and isolated areas. There is an endless series of peculiarities concerning the demographic structure, the size and structure of the local economies, the infrastructural difficulties, etc. which make the adaptation and introduction of the most appropriate innovative elements extremely urgent in order to overcome the negative aspects of peripherality in Europe.

The main argument in this article links the spatial aspects of the emerging new patterns of innovative regional and local development to the ongoing social and economic restructuring. In this way it intends to analyze the background for the formulation and the successful implementation of new types of regional policies which are needed to overcome peripherality in the European Union.

## 4.2 Restructuring, Innovative Growth and Peripherality

In the 1970s and 1980s there were significant shifts in the structure of the economic landscape. Quite dramatic structural changes occurred as the growth of some sectors slowed down or ended, and as new groups of growth sectors emerged. These differences in output growth rates along with differentials in productivity growth rates led to marked changes in the composition of employment. Variations in the competitiveness of economic activities in different regions led to differential rates of regional growth and an increase in spatial inequalities. Temporal instabilities were a cause of rapid changes in the rate and direction of development.

These developments involved persistent differentials in output and productivity growth rates across sectors. As a consequence the shares of different industries in output and employment changed. De-industrialization coexisted with the rise of new growth ensembles (high-technology industries, revitalized craft industries, and producer service activities). In these circumstances the map of economic development was transformed (Dunford and Benko 1991). The productive system transformed as a result of the development of what one can call a new technological trend and a new wave of perpetual innovation. There was an acceleration in the development of new technologies, processes and products and, in particular, a rapid development and convergence of electronics, computing and telecommunications. These new technologies are centred on two principles: the principle of integration which depends on communication and coordination, and the principle of flexibility.

The old productive order centred on the large Fordist firm and mass production, if it has not disappeared, had undergone major changes. With these developments went major changes in work organization and the structure and location of economic activities with the development of new centres of innovation and growth. There was a change in the relationships and allocation of investment funds between industrial, commercial, financial and land and property development activities. In particular industrial activities were subordinated to commercial and marketing functions, a set of producer services, an increasingly independent financial sector and research, development and innovation functions.

These structural changes and the increased role of integrative and management functions resulted in a transformation of job structures. In industrial production, direct production jobs fell relative to maintenance, while in industry 'periproductive' jobs in marketing, research, design, sales and distribution increased relative to production jobs. The growth of internal industrial sector services (which amount to some two-thirds to three-quarters of the services a firm uses) and of marketed services, the high-skill content of these jobs, and the concentration of these activities in major cities where there is a large concentration of professional workers are major factors in the polarization of economic activities. In some ways these diagnoses identified significant trends (Veltz 1991). With the emergence of new growth sectors (in information and communications technologies, in designer industries, and in financial, management, legal and real estate services), and the greater competitiveness of some new industrial zones, one could identify three different types of regions: (1) areas centred on revitalized craft industries, (2) complexes of high technology industries, and (3) metropolises with concentrations of producer services. Some of these activities exhibited strong agglomeration tendencies, and it was in these zones that some of the fastest rates of economic growth were recorded.

One of the phenomena of the 1980s was the development of zones of small and medium sized firms. To what extent did the development of these zones flow from a logic of vertical disintegration or from the establishment of vertical near-integration networks with large firms and research centres. In the early stages of the development of new technologies there was a sharp increase in new firm formation, and some of these new firms emerged as major actors in global markets, while in the late 1970s and 1980s new firm formation remained high. With these developments it was often argued that traditional vertically integrated concerns were disadvantaged compared with more dynamic and flexible industrial systems that emerged through processes of vertical disintegration. Much of the research on French technopoles and on high technology complexes emphasize the central role of vertical disintegration and the development of complex synergies amongst educational and research institutions, complexes of innovative small and medium-sized firms and service companies.

Moreover in so far as differences in growth rates endure the map of regional economic development is remade. At the same time, however, existing distributions exercise a strong inertial effect, new technologies can rejuvenate mature industries, the attractiveness of established centres is far from insignificant, and the scope for selective investment aimed at changing the structure and use of older areas is substantial. In order to explain the spatial logic of the new era there is a need to draw on several of the sets of ideas identified earlier: the sectoral and corporate logic of the new spatial division of labour and models of innovative milieux and geographical industrialization. There is, in short, a dual process: the dynamic and uneven development of a socio-spatial division of labour within large groups and within strategic alliances among major corporations on the one hand, and the development, around the nodes at which these groups locate, of territorial clusters of activities and complex processes of territorial development on the other.

These developments favour locations alongside computer and telecommunications networks and on major road, rail and air transport infrastructures. The particular geographical shape of these developments will depend on the distribution of major infrastructural investments as these investments are the precondition for the implantation of these integrated activities. Included are planned investments in the means of transport and communication required to run organizations at a distance and allow interaction of units and in support services and cultural residential facilities. But the above elements are exactly the missing links for the incorporation of the LFRs into the remaking of the new European territory. The integrative character of the new technologies within the context of the globalization/localization process seems to reproduce peripherality at the same time that transcends spatial inertia and rewards local initiatives and networked cooperation between areas.

## 4.3 The Resurgence of Regions and the Scope for Local Policies

During crisis, the restructuring and reformulation of the relationship between state and society and of the specialized policies themselves, intensifies and becomes a generalized practice characterising all social agents while new forms of regulation become more appropriate to the ongoing transformations of the division of labour between state and society. We are specifically concerned with the idea that parallel to the spatial shifts which accompany the intensification of economic restructuring, there is also an intensive restructuring of the various regulatory functions of state policies effected at the different spatial levels (i.e. local, regional, national and international).

During this process, the opposite poles of the local or regional and the international seem to acquire a new and upgraded significance vis-a-vis the national as the territorial basis of regulation. However, it is the local level that mostly attracts the interest of current research with important practical and theoretical consequences (Getimis and Kafkalas 1992). In particular, it is a series of individual cities and regions which have become the loci of a series of development initiatives that incorporate a vast range of strategies of innovative growth aiming at the increase of the competitiveness of the local economy. These strategies include items as diverse as the provision of real services to producers, the promotion of technology initiatives, the agreements with inward multinational investment, the undertaking of coordinating capacities by the local authorities and the intervention in the fields of the working and communal life (Murray 1992).

The emphasis upon the locality and the association of local systems of production with a new era of capitalist development became quite widespread during the 1980s. In fact the mid-1980s marked a turning point in industrial geography and spatial studies by associating the crisis of a model of development based on mass-production and mass-consumption with the emergence of a new era of flexible specialization where industrial districts of small firms with a communal character become the new dynamic centres of innovation and growth.

It should be accepted that in market economies the examination of regulatory policies should also include the delimitation of the 'non-regulated' areas. For example, the privatized areas of the market either remain 'self-regulated' outside the sphere of state institutions and/or explicitly undermine such formal institutional frameworks. Other non-state traditional institutions of social reproduction, like the church or the family, play also an important role substituting and/or alongside the existing public provisions. Depending on the forms of the social division of labour, only certain areas of the social reality are susceptible to regulation through state action. It is within this context that the fragmentation of policies towards new forms of state intervention, e.g. what might be termed a more 'flexible' state, or towards an upgraded role of the regions and the localities could be systematized (Getimis and Kafkalas 1992).

On the whole it seems that a very important consequence of the competition for innovation based development is the development of contradictions which are expressed with over investment, deterioration of infrastructural provisions and relaxation of labour regulations which though able to provide short term flexibility, might in the long run destabilize the delicate balance of the local labour markets and the importance of efficient transactions for this kind of development (Scott and Storper 1992). To counteract the negative consequences there is scope, in addition to the local and regional policies, to formulate policies and regulations at a higher level than the regional, and to the extent that re-agglomeration and localization tendencies are accompanied by an extreme globalization of competition in financial, product and labour markets, indeed at the European Community level which will promote regional co-ordination and cooperation through networking of regions with a common interest in regulating innovation based development. Existing or remaining national policies should be re-examined accordingly within the context of subsidiarity.

Increasingly, regional and local authorities and representatives of regional interest groups and associations are concerned with the promotion of new institutions and policies which can foster and reproduce the innovative character of their areas. Networking policies should take into account the benefits of horizontal, less hierarchical relations with which firms and large organizations are experimenting and explore the possibility for its expansion to entire regions with a sectoral or functional overlapping of interests.

It is not uncommon for innovative areas like Baden-Wuerttemberg or Emilia Romagna, despite their other differences, to establish specialized institutions for knowledge diffusion, technology transfer and specialized consulting through agencies which link firms with universities and research centres. Such institutional forms can themselves employ an important number of key professionals and/or act as mediators for the increasing mobility of highly skilled labour or the creation of reservoirs of information and specialized services.

Strategies for interconnecting different firms and different activities (research, production, marketing and management) can be linked with spatial strategies. A distinction can be made between strategies of spatial concentration and strategies of spatial disintegration. Several simple models can be distinguished: a model in which key firms are surrounded by a complex network of subcontractors and where the network is characterized by a dense and relatively non-hierarchical kind of vertical near-integration, a model in which production in vertically integrated firms is spread out across the globe to take advantage of differential labour costs and to get access to markets, and a situation that may prevail in Europe where large firms establish automated production near the markets they are serving.

Furthermore, new initiatives have been developing in the area of interregional cooperation and networking, a fact that expands the scope of the above-mentioned processes beyond the specific regional boundaries and towards a European network of regions with a common interest upon the promotion of innovative activities and the transfer of technology. Items of major interest in this respect are the promotion of learning through partnering, the increase of cohesion through cooperation in competition, the diffusion of diverse experiences, and the decentralization of innovation in order to avoid acceleration of polarization tendencies. Cooperative performance itself as a relatively recent area of regional initiatives is under scrutiny, for example through the FAST Four Motors Programme, in order to establish its various aspects.

Traditionally the concerns of regional authorities and the insights of theory emphasize a number of issues which should be taken into account whenever there is need to formulate and implement regional policies or to assess their results. First the creation of local linkages; second, the degree of local content or local value added; third, the skill profile of the work force; and fourth, the extent of autonomy in decision making capacities.

In relation to these issues, until the mid-70s the orthodoxy was that the presence of a high proportion of small and medium sized enterprises was considered to be the weak link of local productive systems, while on the other hand the usual critique of inward investment policies was that the promotion of branch plants by large corporations does not lead to the creation of local linkages and the strengthening of local economies. The logical and political conclusion of such views was that backward regions are rather doomed to a kind of structural underdevelopment unless they find both the necessary finance and the political support to become part of the developed core.

However, these views were turned upside down during the 1980s. First, the existence of small and medium sized enterprises were considered to be a major asset of regions, especially whenever they were part of an extensive system of interfirm cooperation at the local level. In such cases it is possible to create a highly specialized and extremely competitive system area with great adaptability to the volatile markets of the new era. Second, the increase of inward investment, especially in key sectors was particularly welcome in order to help in the creation of employment and the diffusion of key technological and organizational innovations in the host regions.

In this sense in order to understand the prospects of specific areas we should look at both sides; the endogenous potential and its realization through the choices and the initiatives of local actors and the inward investment structure as it is reflected in past and incoming flows. The key to development is the embodiment of all investment into a local productive system. It does not matter whether the plant is part of a global network or a locally integrated system of subcontracting. In fact it is an enormous asset to the area whether all of its establishments are interconnected through a mixture of global and local networks of production and exchange.

## 4.4 Policies for Innovative Growth and the Overcoming of Peripherality

These developments are of major interest to the peripheral regions because they can become competitive players on the basis of their own assets to the extent that they will be able to provide an adequate institutional background for the introduction of innovative forms of growth in the local productive system. This is certainly a possibility that relies upon the existence of and the ability to use the new infrastructural and technological innovations. (Murray 1992; Morgan 1992). It seems that the promotion of both the above conditions, i.e. institutional thickness and the grasp of new technology, are within the range of regional policies. We can outline a number of new policy fields through which significant advancements can be expected in overcoming peripherality.

### 4.4.1 Technology Transfer

The ability of both private enterprise and local administration to induce innovation into the productive system and benefit from the application of innovations produced elsewhere, constitute a central component of competitiveness and accordingly of the relative position of the various regions and firms into the division of labour.

In an era of continuously contracting product life-cycles and accelerating depreciation of basic fixed capital both centrality and peripherality tend to be defined by the conditions of innovation led development. It is very important to understand that the isolated firm and especially the small and medium one could not take on the huge amount of resources needed for R&D in new products and processes.

As a result of the above there is increasing pressure on the public sector and other intermediate bodies (local and regional authorities, universities and research institutions, collective initiatives, etc.) to bear at least part of this cost in order to strengthen the innovative capacity of the local economies. This fact is expressed for example, through the establishment and financial support of academic and research units and the strengthening of their links with the private sector and the local economy in general. Furthermore, there is support to the so called high risk capital investments in new products and processes, the planning of innovative places such as science parks and technopoles and the creation of centres for the transfer and adaptation of new technology (Komninos 1993; Cooke 1993).

If the public sector could not respond effectively the only alternative is its substitution by the productive system itself, i.e. through strategic alliances of enterprises in leading branches or through the spatial agglomeration and networking of numerous smaller enterprises trying to benefit from the external economies and cooperation. In these cases the specificities of the institutional framework and the existing public policies (i.e. concerning procurements) directly affect the innovative potential of the region.

### 4.4.2 Training and Reskilling

It has been widely accepted that a large part of the registered unemployment has a structural character and is due to the social and economic restructuring of the older forms of development observed in the beginning of the article (i.e. de-industrialization), while it is often expressed through skill shortages concerning the new innovative branches and economic activities. The form of regional policies which are expected for the overcoming of structural unemployment and the newly created skill shortages, are linked with the development of methods of initial training and reskilling in the new technological demands of the productive system.

Thus, the appropriate regional policies are expressed through the establishment of training centres and the support of training schemes for on the job training in the enterprises. The expansion and the enlargement of both plans and infrastructures for education and training have certain targets which are linked with the accessibility to the labour market, the increase of labour mobility, the promotion of equal opportunities between the various social categories and geographical areas and finally, as the combined result of the above, the matching of demand and supply in the labour market.

Furthermore, with the promotion of specialized training programmes it becomes possible to implement social policy in a positive discrimination manner in order to benefit particular social groups (i.e. women, youth, handicapped, migrants, etc.) which could then integrate directly into the labour market instead of being beneficiaries of state subsidies and other traditional forms of social policy of a philanthropic nature.

### 4.4.3 Environmental Protection and Sustainable Development

The emerging new forms of development stress the environmental issues because, on the one hand, there is great awareness of the symbiotic relationship between man and nature and on the other hand there is increasing demand for better quality of the physical and built environment, especially in the new spaces of innovative growth where the environmental conditions become a prerequisite for development.

Thus, environmental protection and upgrading becomes a new and important field for regional policies. It is not only that there are vast long term interests on the balance of the ecosystem but the very details of everyday life and well being of the citizens have become environmentally sensitive. Finally, we should not underestimate the fact that the political and economic linkages of the new environmental policies extend towards many directions and time horizons.

The processes which are connected to energy conservation and recycling, the protection of plant and animal species, the restoration of the natural and historical character of concrete areas, the shift of production to ecological products and organic processes, are all at the focus of political and economic conflicts which demand the formulation of institutional frameworks and the implementation of specialized regional policies as the cost and benefit of the environment could not be easily allocated to the various social partners.

## 4.5 Conclusions: Towards a Non-Polarized European Space?

The document Europe 2000 published by the European Commission (1991) deals with the monitoring of information and the scenarios concerning regional planning and the spatial patterns of development in Europe. The document expresses explicit concern over peripheral and especially the border regions of the Community and argues for the need to implement policies and planning measures in cooperation with complementary activities to be undertaken by the national and regional authorities in order to avoid the acute problems of peripherality.

This discussion is placed within the context of the various scenarios for the spatial future of Europe (Kunzmann 1993; Masser et al 1992). The identification of areas like the well known 'blue banana' and the southern sunbelt zone, independently of their relevance, demonstrate both the dynamism of the new growth poles and the reproduction of peripherality at the level of the new European territory. On the other hand there are national and regional concerns for spatial planning and regional development policies which emphasize the strengthening of the national and/or regional territories. The challenge for the European Union is to combine all the above concerns and tendencies within the context of subsidiarity and to promote both the dynamism of innovative growth areas and the introduction of innovative regional policies for the overcoming of peripherality.

## References

Amin A, Thrift N (1993) Living in the global. Paper presented at the Conference on Cohesion and Conflict in the Single Market, 18-20 November 1993. Newcastle upon Tyne

Andrikopoulou E (forthcoming). Regional policy in the European Community. Themelio, Athens (in Greek)

Begg I (1993) Threats to cohesion. Paper presented at the Conference on Cohesion and Conflict in the Single Market 18-20 November 1993. Newcastle upon Tyne

CEC (1991) Europe 2000. Outlook for the development of the Community's territory. Brussels
Cooke Ph (1993) Regional innovation systems. An evaluation of six European cases. In: Getimis P, Kafkalas G (Eds) Urban and regional development in the new Europe. URDP/TOPOS, Athens 193-206
Dunford M, Benko G (1991) Structural change and the spatial organisation of the productive system. An introduction. In: Dunford M, Benko G (Eds) Industrial change and regional development. The transformation of new industrial spaces. Belhaven Press, London 3-23
Dunford M (1992) Socio-economic trajectories, European integration and regional development in the EC. In: Dyker D (Ed) The European economy. Longman, London 155-181
Dunford M, Kafkalas G (1992) The global-local interplay, corporate geographies and spatial development strategies in Europe. In: Dunford M, Kafkalas G (Eds) Cities and regions in the new Europe. Belhaven Press, London 3-38
Getimis P, Kafkalas G (1992) Local development and forms of regulation. Fragmentation and hierarchy of spatial policies in Greece. Geoforum 3:73-83
Holmes P, (1992) The political economy of the European integration process. In: Dyker D (Ed) The European economy. Longman, London 51-69
Ioakimidis PK, (1993) European Political Union. Theory, negotiation, institutions and policies. Themelio, Athens (in Greek)
Kafkalas G, Komninos N (1991) Local development strategies and the new role of state policies. Presented at the seminar Aspects of social policy today, organized by the Garagiorgas Foundation. 27-29 November, Athens
Komninos N (1993) Innovative growth in peripheral regions. Some implications for Greece. In: Getimis P, Kafkalas G (Eds) Urban and regional development in the new Europe. URDP/TOPOS, Athens 193-206
Kunzmann RK, (1993) Defending the national territory. Spatial development policies in Europe in the 90's. Paper presented at the Conference on Cohesion and Conflict in the Single Market. 18-20 November 1993, Newcastle upon Tyne
Maraveyias N, (1993) The Maastricht Treaty and the economic and social cohesion of the European Community. In: Christodoulidis I, Stefanou K (Eds) The Maastricht Treaty. A synthetic approach. Sideris, Athens 261-287 (in Greek)
Masser I, Sviden O, Wegener M (1992), The geography of Europe's futures. Belhaven Press, London
Mitsos A (1993) Post-Maastricht process towards European Union and the economic and social cohesion of the Community. In: Getimis P, Kafkalas G (Eds) Urban and regional development in the new Europe. URDP/TOPOS, Athens 1-6
Morgan K, (1992) Innovating by networking. New models of corporate and regional development. In: Dunford M, Kafkalas G (Eds) Cities and regions in the new Europe. Belhaven Press, London 150-169

Murray R, (1992) Flexible specialisation and development strategy. The relevance for Eastern Europe. In: Ernste H, Meier V (Eds) Regional development and contemporary industrial response. Extending flexible specialisation. Belhaven Press, London 197-218

O'Brien R, (1992) Global financial integration: the end of geography. Pinter Publishers, London

Storper M, Scott A (1992) Regional development reconsidered. In: Ernste H, Meier V (Eds) Regional development and contemporary industrial response. Extending flexible specialisation. Belhaven Press, London 3-24

Storper M, Scott A (1993) The wealth of regions. Market forces and policy imperatives in local and global context. WP 7, The Lewis Center for Regional Policy Studies, UCLA, Los Angeles

Veltz P (1991) New models of production organisation and trends in spatial development. In: Dunford M, Benko G (Eds) Industrial change and regional development. The transformation of new industrial spaces. Belhaven Press London 193-204

# 5 Barriers in Network Performance in Border Areas

M. Giaoutzi and A. Stratigea

## 5.1 Introduction

In the last two decades a great diversity of scientific and technological developments have taken place which have had a considerable impact upon the dynamics of the world economy. The patterns of competition have been dramatically altered, while the internationalization of competition has forced countries to participate in broader economic entities in order to strengthen their position in the emerging new economic order.

In such a context great importance is attached to *New Information and Communication Networks*, since a considerable amount of interaction taking place among the various actors is concerned with information exchange. Access to such networks reinforces the *efficiency* and *competitiveness* of the actors involved, thus constituting a strategic resource in the light of the new era. They remove the negative potential resulting from peripherality and isolation.

The various types of world regions exhibit different structural characteristics which result in place-specific development paths. These are forming diversified socio-economic settings, which can disrupt the smooth flow of information in space having as a result the appearance of *barriers to communication* among different regions or actors of a system. Barriers to communication can influence the *intensity* or even the *existence* of communication channels, with serious implications for the development potential of each region.

Among the various types of regions, *border areas* constitute an interesting type of region, which due to their specific characteristics can have considerable spatial effects in the context of barriers to communication. In recent years, a large amount of research effort has been focused on the issue of border areas, defining them either on a strictly geographical basis or by means of their gateway function between two worlds.

The focus of this paper will be on a systematic presentation of the various approaches encountered in the literature on the subject of barriers to communication in border areas. Emphasis will be placed on the modelling efforts adopted in the context of the spatial impact of border areas.

## 5.2 Barriers to Communication in a Network Context

In our modern information era, communication becomes a key element through which valuable information can be acquired. Any discontinuity in the process of communication may cause critical disadvantages to the regions and actors involved.

A great amount of research has recently been focusing on the study of such discontinuities and their spatial implications, addressed as barriers to communication. The issue of barriers to communication has been approached by many researchers from both a theoretical and a methodological point of view.

*Barriers to communication* may be interpreted as all obstacles in space or time that - apart from normal average distance friction costs in spatial communication - impede the smooth transfer or free movement of information related activities (Nijkamp and Rietveld 1989).

*Communication networks*, in such a context, become of crucial importance in the process of information exchange. From a spatial perspective, a network is considered as a set of interconnected locations or places that, together, make up a system. The system can have one or more functions; it can be simple or complex, monofunctional or multifunctional, depending on the structure of the network and on the types of flows carried (Suarez Villa et al 1992). A network will typically comprise a set of nodes, the links connecting the nodes, and the flows, whose quantity and quality are the ultimate objective of the system (Smith 1976; Gould 1991; Suarez Villa et al 1992). The existence of links connecting the nodes plays a significant role, since they serve not only a short term function of providing the necessary routes along which flows between nodes take place, but also a long term function of shaping the development potential of the nodes themselves. Barriers to communication may stem from a variety of reasons such as *linguistic, cultural, legal, institutional differences* etc. (Nijkamp et al 1990; Rietveld and Janssen 1990).

Very often barriers to communication are related to the *context of national borders*. Rietveld and Janssen (1990) have approached this issue by examining barriers to communication due to the existence of national borders by means of telephone traffic data. In such a framework, barrier effects of borders are estimated in order to capture discontinuities in spatial interaction across political boundaries (Rietveld and Janssen, 1990; Giaoutzi and Stratigea 1990). Moving along the same rationale, Rossera's (1990) work aims at identifying barriers to communication due to political boundaries based on the telex network.

Barriers to communication may also be approached in a *regional context* of analysis (see also Figure 5.1). Such a study of barriers has been undertaken for the case of Switzerland (Rossera 1990), where the empirical results show that interaction patterns are considerably marked by the existence of the four officially recognized ethnic and cultural communities. As Rossera shows, the interaction taking place between the different communities is rather limited (Rossera 1990). Donze's (1990) work along the same line approaches the issue of barriers to communication by identifying the type of barriers existing between Swiss regions and regions in the adjacent countries.

| BARRIERS TO COMMUNICATION | | | |
|---|---|---|---|
| | NATIONAL | REGIONAL | BORDER |
| SPATIAL CONTEXT | Rietveld and Janssen 1990 | Rossera 1990 Donze 1990 | Giaoutzi and Stratigea 1990 |
| | Rossera 1990 | Fischer 1990 | Suarez Villa et al 1992 |
| | Giaoutzi and Stratigea 1990 | Klaassen et al 1972 | |

**Fig. 5.1.** Barriers to communication in a spatial context

Apart from the national and regional scale of analysis of barriers to communication, *the border region context* has also received due attention by a number of researchers and will be thoroughly studied in the present chapter. Border areas may play a special role in a network economy, since empirical evidence supports that they represent a rather important case-specific context of analysis. (see also Figure 5.1)

Barriers to communication can also be identified in the context of *various sectors* such as *trade sector, telecommunications sector, transport sector* etc. (see also Figure 5.2).

Barriers appearing in the *trade sector* are mainly related to:

- *Institutional barriers*, reflecting the interest of national policies in their territories;
- *Technical barriers*, referring to different standards holding in different countries;
- *Social barriers*, referring to the conflicts between different objectives or groups, which prevent a more coherent network configuration;
- *Network barriers*, due to either missing links between the various nodes or actors involved, or missing networks, i.e. absence of strategic layers or components of transport and communication infrastructure (Giaoutzi and Nijkamp 1994).

| BARRIERS TO COMMUNICATION | | | |
|---|---|---|---|
| | TRADE | TELECOMMUNICATIONS | TRANSPORT |
| SECTOR CONTEXT | Rietveld and Janssen 1990 | Capello and Nijkamp 1994 Nijkamp et al 1990 | Salomon and Tsairi 1994 |
| | Rossera 1990 | | |

**Fig. 5.2.** Barriers to communication in a sector context

The identification of barriers appearing in the *telecommunications sector* seems to be of crucial importance, since radical technological innovations occuring during the last two decades have reinforced the role of telecommunications services and infrastructure in the economic system.

Barriers related to the *telecommunications sector* may refer to all kinds of phenomena, i.e. economic, organizational, technological, cultural, political, etc. which impede a smooth penetration, adoption and development of the Information and Communication Technology Sector (ICT) (Capello and Nijkamp 1994). Barriers appearing in such a context may be either *exogenous* to a certain system (e.g. language barriers, physical barriers) or *endogenous* to the system (e.g. congestion due to the limited capacity of a network). According to the point of view adopted - demand or supply side - different types of barriers may appear. Capello and Nijkamp present a list of barriers which are either demand or supply-side oriented (Capello and Nijkamp 1994).

Barriers to communication in the *transport sector* from a user perspective may be summarized as follows (Salomon and Tsairi 1994):

- *Cost barriers*, reflecting the unequal distribution of costs, which result in the distortion of physical distances;
- *Language barriers*, impeding the cooperation between the various actors;
- *Physical barriers*, resulting in the lack of transport infrastructure between the various nodes of a network;
- *Cultural barriers*, which - though difficult to define - are widely accepted to play a very important role in hampering the cooperation among the various actors at the international scene.

Barriers to communication can be either *exogenous* or *endogenous to the system.*

- *Exogenous barriers* are caused by factors external to the system, and may refer to: (Nijkamp et al 1990; Giaoutzi and Stratigea 1990; Suarez Villa et al 1992)
- *Cost barriers*, referring to the costs related to this connection;
- *Language barriers*, applying between sender and receiver;
- *Political and Cultural barriers*, referring to the actors involved;
- *Physical barriers*, resulting either in missing links or in missing networks between the actors involved;
- Different *time zones* among the various actors, which hamper the establishment of a proper connection between them.

*Endogenous barriers*, on the other hand, are caused by factors internal to the system and may refer to (Nijkamp et al 1990; Giaoutzi and Stratigea, 1990; Suarez Villa et al, 1992; Giaoutzi and Nijkamp, 1994):

- Barriers related to *missing links* between sender and receiver, i.e. there is no communication link between them;
- Barriers related to the *capacity* of the network, e.g. congestion which may act negatively on the establishment of communication among the various actors.
- Barriers associated with *limited access* to the network.

The conclusion is thus that there is a wide spectrum of types of barriers.

## 5.3 A Review of Modelling Network Barriers

Borders and barriers are evidently playing a critical but often neglected role in spatial and socio-economic dynamics, while historical experience has proven that the removal of barriers may have a substantial impact upon the growth of regions or nations (Batten and Johansson 1991; Nijkamp 1994).

A large amount of modelling efforts has been oriented towards the issue of barriers hampering an interaction among the various actors in a spatial system which can be classified according to the:

- *scale of analysis*;
- *emphasis placed on link and node barriers*;
- *methodological approach adopted.*

The *scale of analysis* refers to the *macro, meso and micro approaches* mentioned above, implying accordingly the international, national/local and individual levels.

At the *macro/meso* level of analysis, the modelling efforts have often focused upon barriers appearing between regions or countries by using a *gravity type spatial interaction model (SIM)*. In this context, often a modified version of the gravity model has been utilized, namely the log-normal model applied to the unconstrained form of the gravity model. Empirical results are often the output of applying *Ordinary Least Squares (OLS) regression to the log-normal model* (see Figure 5.3).

A number of case studies appear in the literature, dealing with the study of barriers which use a *log-normal gravity model.* Along these lines are the work by Klaassen et al (1972), who studied the implications of language barriers between the Flemish and the Walloons in Belgium; the study by Fischer et al (1990) and (1992), Giaoutzi and Stratigea (1990) and Rietveld and Janssen (1990), who tried to identify the existing barriers in the international telecommunications network; and the study by Rossera (1990 and 1991), focusing upon barriers to communication at the international level, with emphasis on the cultural differences between different nations. For this purpose the latter author used telex and telephone flows for communication.

The same methodological approach - a regression model of the gravity type - has been used for the analysis of barriers in the European Road Network in order to study the impact of borders as barriers in the road network on the accessibility of European cities (Bruinsma and Rietveld 1994) (see Figure 5.3).

The work by Guldman (1993) refers to a *meso* level of analysis employing a different methodological approach than those mentioned above. Guldman creates an *inter-sectoral input-output table* depicting the telecommunication flows among the various sectors of a regional economy. In this methodological approach, which focuses on the understanding of the structure of telecommunications, business demand can also be used to identify the barriers to communication which exist between various sectors at the regional level (Figure 5.3).

Finally, barriers to communication can also be identified at the *micro level* of analysis. An important aspect at this level is referring to the impact of the new electronic communication media on individual behaviour. This relates to the study of barriers, which may affect individual behaviour in the process of choosing communication media for the production and distribution of knowledge and information. Identification of barriers at the micro level is considered of critical importance, since the role attached to the exchange of knowledge and information in the network economy context has been largely recognized as being a prerequisite for the survival and growth of the actors involved.

| | SCALE OF ANALYSIS | | | |
|---|---|---|---|---|
| | Gravity SIM Log-normal | Input-Output Models | SIM Utility Maximization | Random Utility Approaches Discrete Choice |
| **Macro** | Klaassen et al 1972<br>Rossera 1990, 1991<br>Fischer et al 1990, 1992<br>Bruinsma and Rietveld 1994<br>Rietveld and Janssen 1990<br>Giaoutzi and Stratigea 1990 | | | |
| **Meso** | | Guldman 1993 | | |
| **Micro** | | | Domencich and McFadden 1975 | Fischer et al 1992, 1993<br>Ouwersloot 1994 |

**Fig. 5.3.** Modelling efforts according to the scale of analysis

At this level of analysis, choice processes of individuals or households are commonly adressed by using *utility-maximizing models*. Examples of such behavioural approaches to the problem of spatial interaction are found in the work by Domencich and McFadden (1975), Golob and Beckmann (1971) etc. (Figure 5.3)

In more recent research endeavours, the classical utility-maximizing approaches have been overshadowed by *random utility approaches* as well as *discrete choice approaches* in the form of logit, probit and elimination-by-aspect (EBA) models. Application of these methodological tools can be found in the recent work by Fischer et al (1992), Fischer et al (1993) and Ouwersloot (1994), who studied the issue of barriers inherent in the choice of communication media at the individual level, focusing particularly either on barriers to communication within existing networks or on barriers to communication related to network extension (Figure 5.3).

Barriers affecting communication patterns might be introduced in the context of a *gravity type spatial interaction model* in two ways (Batten and Tornqvist 1990):

- via the *origin or destination variables*, when the barrier affects the relative size of the population of candidate senders at the place of origin, or receivers at the place of destination (e.g. budget constraints, accessibility to handsets, exchange conditions etc.), and
- via *the friction interaction variable* - or distance dij -when the barrier is such that the likelihood of interaction traffic between origin and destination (e.g. physical topography, language differences, cultural differences, tariffs, frontiers etc.) is affected.

The first type of barrier can be considered as being node-specific and might be termed as a *node barrier*, while the second is link-specific and might be called a *link barrier*. (see Figure 5.4).

The distinction between node and link barriers has modified the classical gravity approaches, and can be better adapted to the concept of a network, consisting of links and nodes. Node and link barriers introduce important distinctions in the context of both theoretical and analytical approaches. The above distinction has been adopted by Rietveld et al (1991) who studied the factors discouraging communication in international telecommunications network (Rietveld et al 1991).

| | **LINK AND NODE BARRIERS** | |
|---|---|---|
| | **Type of Barrier** | **Gravity type SIM** |
| **Node Barrier** | Barrier affecting the relative size of population of candidate senders/receivers | Rietveld et al 1991 Batten and Tornqvist 1990 |
| **Link Barrier** | Barrier affecting the likelihood of interaction between origin and destination | Rietveld et al 1991 Batten and Tornqvist 1990 |

**Fig. 5.4.** Modelling efforts according to the type of barrier - link and node barrier.

The analysis of barriers to communication by use of a gravity type spatial interaction model -although it offers a certain potential in the context of barrier analysis - still possesses some serious drawbacks. The most important of them are referring to (Batten and Tornqvist 1990):

- its macroscopic and descriptive character, where it identifies only the combined effects of various barriers, which are operating simultaneously, and not the particular contribution of each individual barrier, and
- its inefficiency to cope with network barriers.

In most recent *methodological approaches*, the notion of networks has been very much emphasized during the last decades. In a network approach, spatial development patterns are no longer local and hierarchical, but also international and multi-layered in nature. In studying barriers to communication, it is necessary to understand the multidimensional structure of network systems (Nijkamp et al 1990; Giaoutzi and Stratigea 1990; Batten and Tornqvist 1991). In this context, the issue of *network barriers* discouraging contact between actors belonging to different networks or contact systems appears extremely important. Network barriers are both multidimensional and hierarchical in character. As Batten and Tornqvist state, classical methodological approaches are not sufficient in coping with network barriers, and therefore methodological tools must be employed which are more relevant in a network context. Along these lines the issue of accessibility and interactivity as multidimensional measures of contact potential are issues which deserve more attention in the context of network barriers, and therefore various analytical approaches are proposed for further discussion (see Batten and Tornqvist 1990) (Figure 5.5).

Another approach has been recently proposed by Fischer (1994), based upon a *general nested sigmoid neural network model* (Fischer 1994). The feasibility of the above model has been illustrated in the context of modelling interregional telecommunication traffic in Austria, and its performance has been evaluated in relation to the classical regression approach of the gravity type. The primary advantage of general neural networks in relation to the classical regression approach to spatial interaction modelling lies, according to the author, in the fact that it is capable of handling a more general functional form than the gravity model, while on the other hand they can principally be used to cope properly with problems, where conventional statistical techniques may be inappropriate or cumbersome to use (Figure 5.5).

The above described methodological approaches are useful for the analysis of spatial interaction patterns associated with communication, but also for transport issues; they refer to macro/meso and micro levels of behavioural analysis. Each of the above levels offers scope for a consideration of barrier effects and their influence on interactive behaviour. The choice of a suitable methodology will largely depend on the nature of the barrier effect under study.

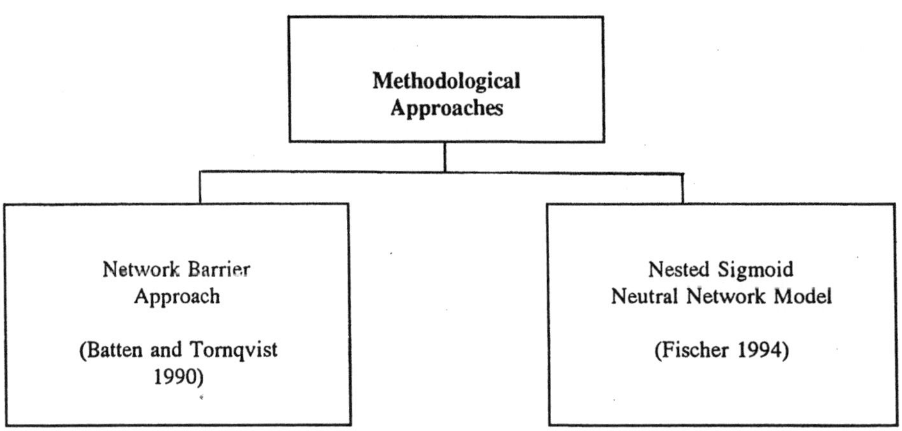


**Fig. 5.5.** Barriers to communication - methodological approaches

## 5.4 Barriers to Communication in Border Areas

Borders were traditionally regarded as barriers to economic development, since they were in general creating impediments which hampered the economic gains of trade (Nijkamp 1994).

In a network economy context it is important to notice the emphasis placed on the concept of borders, which is the result of the growing recognition that border areas should not be viewed any more as peripheral regions but receive due attention for their potential role in spatial developments.

A brief historical review on the border concept gives a broad spectrum of definitions which are subjected by the socio-political developments of the times in each country. *German scholars* thought that borders should encompass homogeneous linguistic and ethnic groups. The *French approach* on the other hand promoted the idea that borders should delimit the 'fatherland' which might encompass a variety of different cultures. According to a *socio-political approach* a border would be a politically negotiated demarcation line, based mainly on multilateral agreements. It must be pointed out however that history provides a few examples which confirm the former statement. Usually the formation of borders has been accompanied by the display of power, either military or political (Hansen 1983).

Various *definitions of border areas* can be found in the literature, with emphasis on both the specific focus of research and the characteristics of the study area.

Nijkamp in his work defines border areas as *peripheral areas* - not necessarily lagging or underdeveloped - which are less oriented towards the center and more so to the external economic world (Nijkamp 1987). Giaoutzi considers border regions as countries at the boundaries of the developed world, which also have the role of a *'gateway function'* between developing and developed countries (Giaoutzi 1988). In the European Context, the distinction between *internal* and *external border regions* has also been introduced, differentiating thus the frontier between two EC members and a EC member with a non-member respectively (Armstrong and Taylor 1993; Hitiris 1991; Geenhuizen et al 1994). Ratti (1990) presents two different views of border areas: the first referring to the *'separation line'* between different nations (political boundaries), and the second one to the view of a *'limiting frontier'* which supports the idea that the border is less seen as a boundary and more as an external limitation, having the characteristic of being mobile. Reichman on the other hand defines border areas as regions distinguished by a certain type of discontinuity in terms of their specific geographical characteristics (Reichman 1989; Ratti and Reichman 1993). Others study border areas as being equivalent to frontiers between states (Anderson 1983; Boot and Van der Veen 1994; Corvers and Hassink 1994).

Border areas can also be perceived as *barrier networks*, with either *closure* or *contiguity* characteristics as their focus of definition (Nijkamp 1987; Giaoutzi 1988) Barriers to communication in a border area context should then focus on these two main axes, viz. closure and contiguity characteristics, the former referring to the type of barrier network (e.g. island, periphery), and the latter to the type and range of barriers involved in each barrier network (Giaoutzi and Stratigea 1990; Suarez-Villa et al 1992).

### 5.4.1 Theoretical Approaches

The issue of borders is only a recent research topic in the regional science research field. In the past borders were mainly regarded as having simply either a negative or positive influence on the economic development of a region. As of the mid seventies Niles Hansen (1977) first started to pay some theoretical attention to the border issue while others soon followed. Hansen in his early work presents the views by Christaller and Loesch, where Christaller considers the separation principle introduced by the borders as being in conflict with rationality and economic principles, while Loesch claims that the goals of the economic landscape - in order of priority - are prosperity, 'kultur' and continuance, although the same goals in the political landscape have completely the reverse order.

In this context Loesch is referring to a set of 'problems' caused by borders, which in reality corresponded to the so-called 'barriers' of today. These are: tariffs which separate economically complementary landscapes, customs, linguistic differences, national characteristics, public contracts and 'official trade' which does not cross the border, as well as the fact that border areas are the most threatened in military terms (Hansen 1977). The overall picture though, presented by Loesch, on the influence of borders on economic development was rather negative.

Christaller however noted that the development of trade economics near borders can be seen as a positive influence of borders, although it has to be admitted that Loesch also mentioned that certain economic activities related to the avoidance of customs duties can have positive effects on border regions (Hansen 1977).

More recent approaches on border areas can be found along two diverse lines of views by Ratti (Ratti 1993, 1994):

- the *border area view*, and
- the *limiting frontier view*.

The *traditional view* is referring to a border area as the territory adjacent to a fixed frontier line - in most cases institutional - wherein the socio-economic effects resulting by the existence of a border are significantly felt. The concept of a border area by Hansen (1977) is clear and rather operational: "This relates to that part of a natural territory in which economic and social life is directly and significantly influenced by the proximity of an international frontier. We thus consider here only open or potentially open regions, excluding regions with a closed natural border situation, like, for example, some parts of the Alps".

In this context a number of relevant concepts appear: the *'border area'* which extends along the boundaries on a national territory expressing in itself the 'centripetal forces'; the *'transbordering area'* which extends across the political boundary and represents the 'centrifugal forces' characterizing the free economy (Ratti 1993); the *'regional intermediate'* presented by House (1981), which refers to the intermediate area between the border area and the national centres. This concept coincides roughly with the activity area of the commuters. The above concepts will be used to present a set of approaches under the umbrella of the *border area* view as follows (Ratti 1993, 1994):

- *the functional activity approach*
- *the core-periphery approach*
- *the regional systems approach*, and
- *the strategic planning approach*

These four approaches will now successively be discussed. The *functional activity* approach places emphasis upon identifying and measuring the functions of a frontier which affect the development of a border area (Biucchi and Gaudard 1981), in three ways:

- first, by estimating the negative *impact* of the frontier *on the flows* (barrier paradigm);
- second, by identifying the *discriminating effects* (positive or negative) which result in differential incomes (e.g., smuggling); and
- third, by ascertaining same *polarising effects* such as location of a specific border or facilitated activities.

The present approach is using the border effects in three ways:

- the *border as barrier* which is inducing its penalising and discriminating effects in the areas adjacent to the border; it is the result of their *peripheral location* versus the centre of the country and the *separation principle*.
- the *'border' in its filtering role* acts as a discriminating mediator between two or more political-institutional and economic systems (House 1981). In economic space a border is the place where the marginal costs equal the price of the product (Di Tella 1982). The introduction of an institutional barrier acts as a filter which modifies the economic space and results in differential rent (Ratti 1993), namely economic conditions for one or the other side of the border where the total outcome does not add up to zero.
- the idea of the *open border* where the function of contact is predominant (Courlet 1988). In this view the economic development of the border regions will no longer be determined by the political-institutional deferential, but by the combination of the comparative advantages of the border areas located at both sides of the borders. The 'open borders' view (Ratti and Baggi 1990) implies a transition from the economic concept of the bordering areas (Hansen 1983) to that of the transbordering economies. This has been the rationale behind the goal of a borderless Europe.

The *'core-periphery approach'* is based on the core-periphery theory, which implies a dualistic spatial segmentation into a centre and an outskirt. In many cases the border areas are not only outskirts in an institutional but also in an economic sense, (Giaoutzi and Stratigea 1989). The complex intra- and inter-regional relationships provoke the need for proper insight into the problem of barriers which would enable the identification of the processes conditioning the structural transformations in the border regions. Therefore, a systematic orientation to the core-periphery problem which would also include the time element would enhance the potential of the approach.

The *regional systemic approach* refers to a systemic interpretation of regional development patterns, which can be considered as a spatial response to changes taking place in a dynamic but contradictory economic system (Stilwell 1991). Borders in such a context may be identified as specific elements of a regional restructuring process. The systemic approach has the advantage to perceive these elements in their functional perspective where the emphasis is placed on the transbordering relationships of the borders which may arise from the encounter of diverse structures (Ratti 1993).

The *strategic planning approach* is placing emphasis on the role of the borders for the individual subjective perception of economic agents and particularly on their strategic behaviour (Covin and Slevin 1989). "The emphasis is upon their surrounding perception and upon their action space in each specific case region. However, people do not act only following a subjective image of the space but also following the function of other external parameters, namely the economic situation, the internal and external politics of the bordering countries etc." (Reichman 1989).

Over the years, these patterns have covered a rather diversified range of structural transformations which to a large extent reflect the diversity and functional multiplicity of their agents and processes (Giaoutzi et al 1990). This calls for an integration of micro and macro approaches which will permit the study.... " of the subjective perception, like other individual actions, which can not be aggregated to a unique model of general access" (Leimgruber 1987).

After the discussion of the border area views, we will now turn to the second concept, viz. the *limiting frontier*. The view of the *'limiting frontier'* is referring to a border as a demarcation line but also as a mobile external limit. Under a political-economic profile the 'limiting frontier' approach concerns the study of barriers to communication with the objective of spatio-economic integration (Peschel 1982).

Four theoretical lines - *in the 'limiting frontier' view* - have been suggested in order to cope with the spatial impact of barriers to communication (Ratti 1993):

- *the spatial micro-economic approach*,
- *the international trade approach*,
- *the spatial individual perception approach*, and
- *the spatial diffusion of innovations*

The *spatial micro-economic approach* (Hansen 1977) refers to the work by August Loesch which gives one of the most interesting presentations of the border effect on the spatio-economic organisation. He distinguishes between political and economic borders which in rather diverse ways have the effect of interrupting the net of market areas (Ratti 1993). The political borders have the role of interrupting the economic space, while the economic borders are defined as the place where the marginal cost equals the selling price. This border is determined by the distance, but above all by other barriers to communication (Ratti 1993).

Another classical author of the same school, Walter Christaller, claimed that the existence of barriers has the inconvenience of limiting the expansion area of commercialized services from central locations. However, the way Christaller is introducing the element of 'time' is interesting, as he observes the way historical incidences condition the subsequent spatial organization. Both authors recognize that the negative effects - due to the existence of barriers - will determine the spatial effects of distortion and non-integration (Nijkamp 1994).

In the second approach, the *theory of international trade* (Broecker 1984), the study of communication barriers plays an important role, since the question arising, in this respect refers to the effects of economic integration on regional disparity at an aggregate level.

The third approach relates to the analysis of the *spatial individual perception* (Hurriot and Perreur 1990). The concept of border, in this 'limiting view', has centered around the context of the 'geography of perception'. Man perceives a specific territory, and there the border plays a role by the perception a man has about space. Therefore, 'mental maps' can be an important tool for the study of the effects of the border on the perception level, but also on the regional identity level.

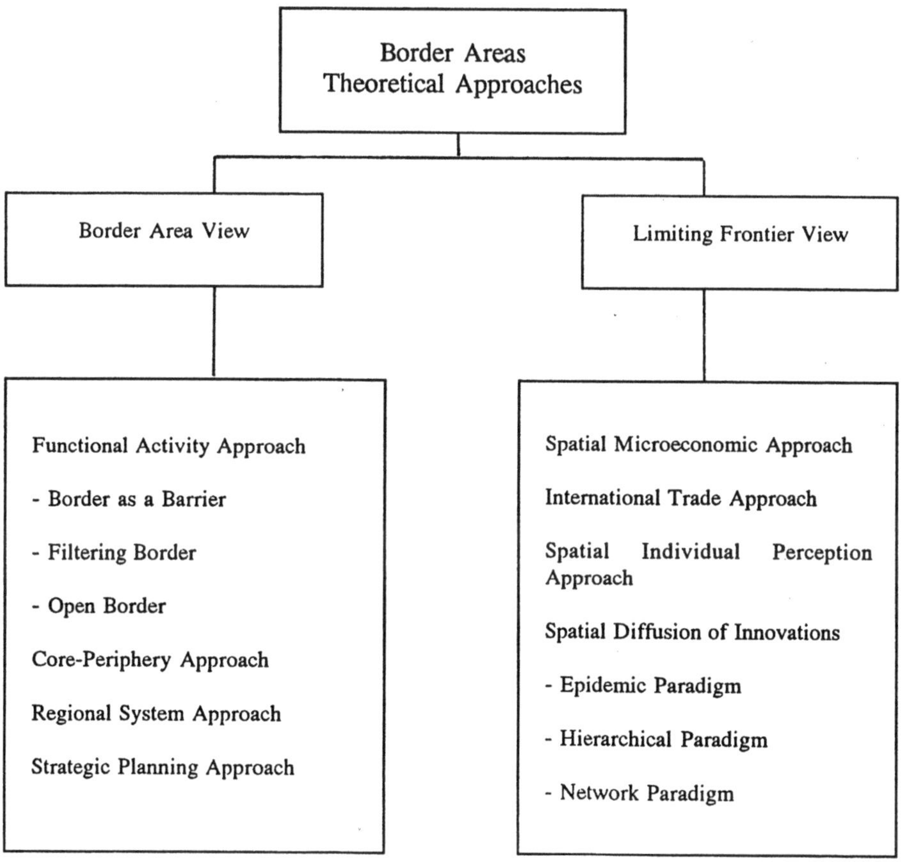


**Fig. 5.6.** Theoretical approaches on border areas (Ratti 1990)

The fourth approach to the study of the spatial impact of communication barriers is the one developed by the theories of *spatial diffusions of innovations* (Brown 1981). The main paradigms developed in this context are: the *epidemic paradigm*, the *hierarchical and the network paradigm*.

The *epidemic paradigm* sees the diffusion of innovation taking place by contagion as a function of communication channels and inversely related to obstacles like distance, economic, sociological, cultural etc.

In the *hierarchical paradigm* the diffusion of innovations is not taking place as a decreasing function of distance but follows more precisely the urban hierarchy. The theory has been introduced mainly by geographers like Christaller and Hägerstrand, but also economists have contributed by offering arguments based on the *theory of 'filtering down'*. Camagni and Cappelin (1984) asserted that the role of the urban media is that of limiting the uncertainty level because of the important economies of the agglomeration.

The *network paradigm* is placing emphasis on the spatial differentiation in the selection environment as an important aspect of the origin and diffusion of new ideas and innovations. This is to a large extent determined by the network freedom and participation of actors. "Network cooperation between actors creates a synergetic surplus. Those actors who succeed to dominate other actors or even entire network segments will consume the synergetic surplus at the cost of others.... Distribution of power over the participants of a network and the ability to monopolize strategic information in a network are important for the diffusion of innovations and for the related distribution of incomes generated" (Kamann and Nijkamp 1991).

The analysis on the diffusion of innovations and therefore, the analysis of barriers to communications, is important for resolving the distortions and disparities in their spatial perspective.

### 5.4.2 Modelling Barriers to Communication in Border Areas

Border region barriers promote the fragmentation of market areas, along with a duplication of services, resulting in contiguity diseconomies that reduce development potential and efficiency (Suarez Villa et al 1992).

Border areas provide the richest context to study network barriers since they comprise the most significant spatial discontinuities. Nevertheless, the scientific literature on modelling barriers to communication in border areas does not appear particularly rich. Most of the modelling and empirical studies presented below consist of a limited sample of efforts although they gradually attract more scientific interest in this research (see also Figure 5.7).

There is a number of empirical models on *economic integration in Europe* which mainly employ gravitational models. Among the most complete models are those using 'regional accessibility' as a growth factor (Keeble 1982) following the traditional expression:

$$P_i = \sum_{j=1}^{n} \frac{M_i}{D_{ij}} + F$$

where $P_i$ = potential of region i
$M_i$ = activity volume in region J
$D_{ij}$ = global transportation costs between region i and j
$F$ = customs taxes in the case of a border between region i and region j.

In this case the effect of a barrier modification is simply reflected by the variation of a customs tax. On the basis of such a gravity model the theory foresees, at an aggregate level, usually a reinforcement of a central area with respect to peripheral areas (Ratti 1993).

Among the studies on border regions of an *international trade* character Herman et al (1982) adopt a regression analysis in order to quantify the influence of cultural factors and compare them to those bound to distance. The central hypothesis is the high weight attached to communication costs, which exist both at the supply (knowledge of markets) and at the demand level (consumer information) and, as Maggi (1989) indicated, which are supposed to be lower as the cultural and lingual proximity is closer. Jeanneret (1985) on the other hand in his study on interchanges of industrial goods and services among the different lingual Swiss regions and the bordering countries, also found the existence of the cultural and lingual proximity effect suggested by Maggi.

A study on the *evolution of territorial organization* in the Canton Ticino by Ratti and Stefano (1986) has explored how this area evolved from distinctly peripheral to an intermediate area where transborder relationships have resulted from the encounter of diverse regional structures.

The model by Rossera on the border effect in the analysis of telephone flows is trying to evaluate systematically the distinctive features of the incorporation of *border regions in the context of interregional communications*. This includes intra- and interregional communications as well as between countries communications where the contacts are classified as short, medium and long distance. The analysis has been carried out by using gravitational models modified with the addition of suitable dichotomous variables for measuring particular effects (Rossera 1990).

There is a number of empirical studies in the context of the *spatial individual perception approaches* where borders are a central notion and enter in strict relation in the mental realm of man' s perception about space. For example, Leimgruber (1980) proposes a useful differentiation on a hierarchical level of three spaces: the first level is that of the daily action space (the space that is immediately perceived); the second is the space of subjective perception (built around a system of values and information); and the third is a space of processes, wherein more objectively the socioeconomic and political processes are functioning. There we find a confrontation between this third space (objective) and the individual space (subjective) where the economist can find an ideal approach for measuring and interpreting the effective role of certain barriers to communication. However the methodological approach still needs to be further developed.

| | Gravity | Regression Analysis | Conceptual Modelling | Statistical Analysis |
|---|---|---|---|---|
| **Economic Integration** | Regional Accessibility Keeble 1982 | | | |
| **International Trade Approach** | | Herman et al 1982<br>Maggi 1989<br>Jeanneret 1985 | | |
| **Spatial Individual Approach** | | | Leimbruger 1980 | |
| **Interregional/ International Communications** | Rossera 1993<br>Bruinsma-Rietveld 1994<br>Rietveld et al 1990<br>Giaoutzi-Stratigea 1990 | Rossera 1990, 1991<br>Klaassen 1972 Fischer et al 1990 | | |
| **Empirical Studies** | | | | Ratti-di Stefano 1986<br>Giaoutzi-Kamann 1993 |

**Fig. 5.7.** Modelling barriers to communication in border areas.

## 5.5 Prospects

The recent theoretical developments on spatial dynamics have given rise to the recognition that spatial development processes are shaped through a complex interplay of both a heterogeneous physical space and a multilayer socioeconomic stratum (Giaoutzi et al 1987, Nijkamp 1984). As a consequence, the multilayer network formation with both interdependencies and hierarchies at all spatial levels, which has become the new model of western economies, has been the methodological context through which the impact of barriers and borders on spatial developments can be approached.

The study of the spatial impact of communication barriers means an orientation towards a process of spatial diffusion, defined as the propagation phenomena in time and space of a specific item - object, institution, idea, practice - of individual groups or any other precisely localised, potentially receiving entity. (Ratti 1993)

For communication barriers and borders it is necesary to identify obstacles - apart from normal distance frictions - that impede the uninterrupted transfer of information, and that result in discontinuity, intensity change and direction of flows.

In such a context the systemic framework is becoming of great importance, while modelling efforts of an integrated nature may prove promising paths towards theorising on border areas development.

## References


Anderson M (1983) The Political Problems of Frontier Regions. In: Anderson M (Ed) Frontier Regions in Western Europe. Frank Cass London

Armstrong H, Taylor J (1993) Regional Economics and Policy Hertfordshire. Harvester Wheatsheaf

Batten D, Tornqvist G (1990) Barriers and Networks. The Methodological Challenge. NECTAR

Batten DF, Johansson B (1991) Origins of the Network Economy. Paper presented at the 31st European Congress of the RSA Lisbon

Biucchi BM, Gaudard G (Eds) (1981) Regions Frontalieres. Georgi, St Saphorin, Switzerland

Boot DT, Veen A van der (1994) Gross Border Cooperation and European Regional Policy. Paper presented at the 34th European Congress of Regional Science Association. Groningen

Broecker J (1984) How Do International Trade Barriers Affect International Trade? Regional and Industrial Development Theories Models and Empirical Evidence. Elsevier Amsterdam 134-147

Brown L (1981) Innovation Diffusion. Menthuen, London and New York

Bruinsma F, Rietveld P (1994) Borders as Barriers in the European Road Network. A Case Study on the Accessibility of Urban Agglomerations. In: Nijkamp P (Ed) New Borders and Old Barriers in Spatial Development. Avebury, Aldershot 139-152

Camagni R, Cappellin R (1984) Cambianto Technologico e Diffusione Territoriale. F Angeli, Milano

Capello R, Nijkamp P (1994) Borders and Barriers in Telecommunication Systems. In: Nijkamp P (Ed) New Borders and Old Barriers in Spatial Development. Avebury Aldershot 209-227

Christaller W (1933) Die zentralen Orte in Süddeutschland. Jena

Corvers F, Hassink R (1994) Why Technology Networks should be stimulated in Border Regions and How. Paper presented at the 34th European Congress of the Regional Science Association. Groningen

Courlet C (1988) La Frontiere: Couture ou Coupure? Economie et Humanisme 301:5-12

Covin JG Slevin PP (1989) Strategic Management of Small Firms in Hostile and Benign Environments. Strategic Management Journal 10:75-87

Di Tella G (1982) The Economics of the Frontier. In: Kindleberger GP, Di Tella G (Eds) Economics in the Long View Vol 1. Models and Methodology. Macmillan, London and Basingstoke

Domencich T, McFadden D (1975) Urban Travel Demand. A Behavioural Analysis. North Holland, Amsterdam

Donze L (1990) Analyse des Flux Telephoniques Suisses. Une Estimation d' un Modele Gravitaire. Institute for Economic and Social Sciences, Working Paper 204

Fischer M, Maggi R, Rammer C (1990) Context Specific Media Choice and Barriers to Communication in Universities. The Annals of Regional Science 24:253-270

Fischer M, Maggi R, Rammer C (1992) Stated Preference Models of Contact Decision Behaviour. In: Academia Papers in Regional Science 71:359-371

Fischer M, Maggi R, Rammer C (1993) Scholarly Communication in Europe. Stated Communication Media Choice and Contact Decision Models Based on Laboratory Choice Experiments in Universities. In: Nijkamp P (Ed) Europe on the Move. Avebury, Aldershot 311-329

Fischer M (1994) Artificial Networks Neurocomputing A Low Dimensional Real World Application Telecommunication Flows Interregional Neural Net Telecommunication Model Preferences for JEL Classification. Austria

Giaoutzi M, Nijkamp P (1988) Informatics and Regional Development. Avebury, Aldershot

Giaoutzi M (1988) Spatial Patterns of Information Technology. The Case of Greece. In: Giaoutzi M, Nijkamp P (Eds) Informatics and Regional Development. Aldershot, Gower 235-248

Giaoutzi M, Stratigea A (1989) Barriers to Communication. The Case of Border Areas. Paper presented at the 29th European RSA Congress, Cambridge

Giaoutzi M, Stratigea A (1990) Telephone Calls and Communication Barriers in Greece. National Technical University Athens. Paper presented at the 32th RSA Meeting, Brussels

Giaoutzi M, Kamann DJ (1993) The Role of Corporate Structures in Border Regions. In: Ratti R, Reichman S (Eds) Theory and Practice of Transborder Cooperation. Bellinzona Press Switzerland

Giaoutzi M, Nijkamp P (1994) Barriers and Missing Networks in European Infrastructure Inland Waterways and Coastal Transport. In: Nijkamp P (Ed) New Borders and Old Barriers in Spatial Development. Avebury, Aldershot 177-192

Golob TF, Beckmann MJ (1971) A Utility Model for Travel Forecasting. Geographical Analysis 16:74-94

Gould P (1991) Dynamic Structures of Geographic Space. In: Brunn SD, Leinbach TR (Eds) Collapsing Space and Time. Geographic Aspects of Communications and Information. Harper Collins Academic, London 3-30

Geenhuizen M, Knaap B van der, Nijkamp P (1994) Transborder Networking. Shifts in Corporate Strategy. Paper presented at the 34th European Conference of the RSA, Groningen

Guldman JM (1993) Input-Output Analysis of Regional Telecommunication Flows Information Economics and Policy. North-Holland 311-329

Hansen N (1977a) Border Regions. A Critique of Spatial Theory and a European Case Study. Annals of Regional Science 11:1-14

Hansen N (1977b) The Economic Development of Border Regions. Growth and Change 8:2-8

Hansen N (1983) International Cooperation in Border Regions. An Overview and a Research Agenda. International Regional Science Review 8:225-270

Hermann H et al (1982) Kommunikationskosten und Internationaler Handel. Florentz, Munich, FRG

Hitiris T (1991) European Community Economics Hertfordshire. Harvester Wheatsheaf

House JW (1981) Frontier Zone. A Conceptual Problem for Policy Makers. International Regional Science Review 1:456-479

Huriot JM, Perreur J (1990) Distance Espace et Representations. Revue D' Economie Regionale et Urbaine 2:197-237

Jeanneret Ph (1985) Regions et Frontieres Internationales. EDES, Neuchatel, Switzerland

Kamann D, Nijkamp P (1991) Technogenesis. Origin and Diffusion in a Turbulent Environment. In: Nakicenovic N, Grueber A (Eds) Diffusion of Technology and Social Behaviour. Springer-Verlag, Berlin 198-222

Keeble D PL Owens and C Thomson (1982) Regional Accessibility and Economic Potential in the European Community. Regional Studies 16:419-432

Klaassen L, Wagener S, Weg A van der (1972) Measuring Psychological Distance between Flemings and the Walloons. Journal of Regional Science Association 2:45-62

Leimgruber W (1980) Die Grenze als Forschungsobjekt der Geographie. Regio Basiliensis 21:67-68

Machulp K (1980) Knowledge. Its Creation Distribution and Economic Significance vol 1. Knowledge and Knowledge Production. Princeton University Press, Princeton, New Jersey

Maggi R (1989) Towards an Economic Theory of Barriers to Communication. Papers of the Regional Science Association 66:131-142

Nijkamp P (1987) The Strategic Role of High Tech Firms in Peripheral and Border Areas. Research Paper, Dept of Economics, Free University Amsterdam

Nijkamp P, Rietveld P (1989) Barriers to Communication. Conceptual Issues. Paper presented at the NECTAR Meeting on Barriers to Communication, Zurich

Nijkamp P, Rietveld P, Salomon I (1990) Barriers in Spatial Interaction and Communications. A Conceptual Exploration. In: The Annals of Regional Science 24:237-252

Nijkamp P (1994) Borders and Barriers. Bottlenecks or Potential? A Prologue. In: Nijkamp P (Ed) New Borders and Old Barriers in Spatial Development. Avebury, Aldershot

Nijkamp P (1994) New Borders and Old Barriers in Spatial Development. Avebury Aldershot

Ouwersloot H (1994) Information and Communication from an Economic Perspective. PhD Thesis, Free University, Amsterdam

Peschel K (1982) International Trade Integration and Industrial Location. Regional Science and Urban Economics. 12:247-269

Ratti R and Di Stefano A (1986) L' Innovation Technologique au Tessin. In: Aydalot P (Ed) Innovative Environments in Europe. GREMI, Paris, France

Ratti R (1990) The Study of the Spatial Effects of the Borders. An Overview of Different Approaches. ESF-NECTAR, Athens Workshop

Ratti R, Baggi M (1990) Strategies to Overcome Barriers. Theoretical Elements and Empirical Evidence. Working Paper, Workshop, NECTAR, Lund

Ratti R, Reichman S (1993) Spatial Effects of Borders. In: Nijkamp P (Ed) Europe on the Move. Avebury Aldershot UK 115-138

Ratti R (1993) Spatial and Economic Effects of Frontiers. Overview of Traditional and New Approaches and Theories of Border Area Development. In: Ratti R, Reichman S (Eds) Theory and Practise of Transborder Cooperation. Bellinzona Press, Switzerland

Ratti R (1994) Spatial Effects of Frontiers: Overview of Different Approaches and Theories of Border Region Development. In: Nijkamp P (Ed) New Borders and Old Barriers in Spatial Development. Avebury, Aldershot

Reichman S (1989) Barriers and Strategic Planning. A Tentative Research Formulation. Paper presented at the NECTAR Meeting on Borders to Communication, Zurich

Rietveld P, Janssen L (1990) Telephone Calls and Communication Barriers. The Case of the Netherlands. In: The Annals of regional Science 24:307-318

Rietveld P, Lierop J van, Ouwersloot H (1991) Barriers to International Communication. What are Your Ten Most Frequently Called Countries? Free University of Amsterdam

Rossera F (1990) Barriers in International Communications. The Case of the Telex Network. NECTAR Meeting, Lund

Rossera F (1990) Cultural Differences as Structuring Factor in International Contacts. An Analysis of the Telex Communication Flows. Research Notes. IRE, Bellinzona

Rossera F (1990) Discontinuities in Communications among Communities of Different Language in Switzerland. An Analysis based on Data of Long Distance Telephone Calls. NECTAR, Athens Workshop

Rossera F (1991) What Insights can be gained from the Available Data on Regional Telecommunications Connections in Europe? An Analysis of a Publication by the International Telecommunications Union. Draft Paper, IRE, Bellinzona

Salomon I, Tsairi B (1994) Barriers and Communication Technologies in a Global Society. An Overview and a Case Study. In: Nijkamp P (Ed) New Borders and Old Barriers in Spatial Development. Avebury, Aldershot

Smith CA (1976) Regional Analysis. New York Academic Press

Stillwell F (1991) Regional Economic Development. An Analytical Framework. Revue D' Economie Regionale et Urbaine 14:107-115

Suarez Villa L, Giaoutzi M, Stratigea A (1992) Territorial and Border Barriers in Information and Communication Networks. A Conceptual Exploration. In: Tijdschrift voor Economische en Sociale Geografie (Journal of Economic and Social Geography) 2. Amsterdam

# 6 How to Overcome Barriers and Border Effects: Theoretical Elements

R. Ratti

## 6.1 Introduction

On the recent issue of European integration two main topics are usually discussed: 'Europe without frontiers' and 'Europe of regions.

The first one stresses common policies to decrease a certain type of frontiers (the institutional kind) and policies to implement the four freedoms, which will certainly and in general transform the nature of the frontier itself.

The second one, is more global and complex; if it also stresses decreasing certain frontiers, it will tolerate at the same time other geographical, social, cultural and structural limits.

As a result, in 20th century Europe the most important difference from the 'old' Europe of Countries is that it will not consist only in the abolition of barriers and frontiers but also and more generally speaking in the transformation that will accrue to the frontier from a barrier into a contact line.In this context we have to put forward in evidence the theoretical hypothesis that stresses the overcoming of barriers through the constitution of contact areas, allowing interregional cooperation.

In this paper we want to give a synthesis of reflexions conducted within a special NECTAR working group, the Network for European Communications and Transport Activities Research[1]. Two different approaches were developed to overcome barriers and border effects:

- the first, the microeconomic one, studies the frontier through the analysis of the economic actor's strategic behaviour, and is based on the theory of industrial organization;
- the second one, has a mesoeconomic character, and considers the role of the frontier inside a specific supporting space or 'milieu'.

---

[1] For an extended version of the article, see Ratti R, Reichman S (Eds) (1993) Theory and Practice of Transborder Cooperation. Helbing & Lichtenhahn, Basel und Frankfurt am Main

## 6.2 The Overcoming of the Barriers by the Operators Strategic Behaviour

The study of the actors' behaviour facing barriers or border effects can be theoretically analyzed through the modern theory of industrial organisation: in fact, the hypothesis of a *situation of uncertainty, bounded rationality and market failure* connected with the existence of a frontier, corresponds to the thesis of a response in terms of organization.

The capacity to and difficulty or impossibility of organizing a strategic answer to barrier and border obstacles will explain the three *barrier effects, filter effects, or polarizing effects* previously described, and above all, their dynamics. This seems very important *in the era of a Europe without boundaries*, which will create other rapid changes (in terms of decadence or emergency) in border regions.

Let us summarize some necessary theoretical elements:

### 6.2.1 Postulates

We postulate the following statements:

- Besides the 'firm' and the 'market', there exists a third form of organization and regulation of production: the cooperation between enterprises (Richardson 1972: "The dichotomy between firm and market, between direct and spontaneous coordination, is misleading; it ignores the institutional fact of interfirm cooperation and assume away the distinct method of coordination that this can provide").
- The existence of 'uncertainty' requires new functions, particularly of coordination (Camagni 1991: "The presence of unescapable static and dynamic uncertainty in the real world implies the presence of extra costs and therefore new functions to cope with these costs and new 'operators or institutions' organizing these functions and shaping factual behavior").
- The concept of 'transaction cost' constitutes the discriminating instrument between the functions assumed by the firm, the market or the cooperation (Coase 1937: "A firm will expand until the cost of organizing an extra-transaction within the firm becomes equal to the cost of carrying out the same transaction by means of an exchange on the open market or the cost of organizing within another firm").

### 6.2.2 Thesis

The existence of market distortions and transaction costs constrains the enterprise to adopt two types of strategic behaviour:

- vertical integration, following the classical reasoning (Williamson 1975, 1985, 1986);
- horizontal integration, as an intermediary organization form, more flexible and based upon a logic of dialogue (Porter 1980, 1986), particularly attractive for

the SMEs (Small and Medium Enterprises) searching for 'synergies' (Kamann and Strijker In: Camagni 1991).

There is a good probability that these intermediate forms of cooperation will be territory dependent, because of the need to perform following a common code of behaviour (Christensen 1988: "The distance of dialogue is restricted by the need of common code and the need of face-to-face contacts ..... This need for proximity is based on the assumption that common code is embedded in a contextual macroframe characterized by common language, law, value, social background and joint ability of orientation").

### 6.2.3 Corollaries

The existence of a 'border' or any other form of 'barrier to communication' constitutes immediately a determinant factor of uncertainty and costs of transaction. The study of these 'borders' or 'barriers' then becomes a crucial element in the constitution and dynamic explanation of an intermediary organization of the 'reseau (network)' type. (See Figure 6.1)

- The degree of integration of a firm, constituted by the agreement with other enterprises (from the simple licence agreement, to the joint venture to a branch) is a function of the transaction costs or the market access costs (CT-CT or CT'-CT'), or of the control costs of the developed interfirm organization (CC-CC or CC'-CC') and also of a preference curve to the integration, corresponding to the demand curve (D-D or D'-D').
- The transaction costs are an exogenous constraining data. Below a certain level (CT'-CT'), the firm will have the interest to further develop its internal organization. Beyond that point, an inter-enterprise cooperation would constitute a less expensive option in relation to the market solution.
- The control costs are represented by a growing function of the 'integration quantity'. The points E1 and E2 are situations where there are equivalences between the possibility of recurring to the market and that of organizing certain functions of production with other enterprises.

- The optimum is established by considering the demand curve, that is, the preference curve of the integration. The point E3 is lower than E1 because it is possible to account for a certain fear due to the risks of all external cooperation formulas.
- The presence of a border or a barrier signifies a line of high transaction costs (CT'-CT'). It will then allow some intermediary solutions with an integration as high as the market distortion. All uncertainty reductions due to the barrier-border would approach market solutions. This would indicate that the cooperation solutions are not necessarily stable or better.
- The opportunity of the firm to take advantage of a planned solution with other enterprises consists in the lowering of the curve CC-CC toward CC'-CC', by technological innovation and management. This shows the innovative character of the network, but does not necessarily guarantee that it will be advantageous to the firm in absolute terms.

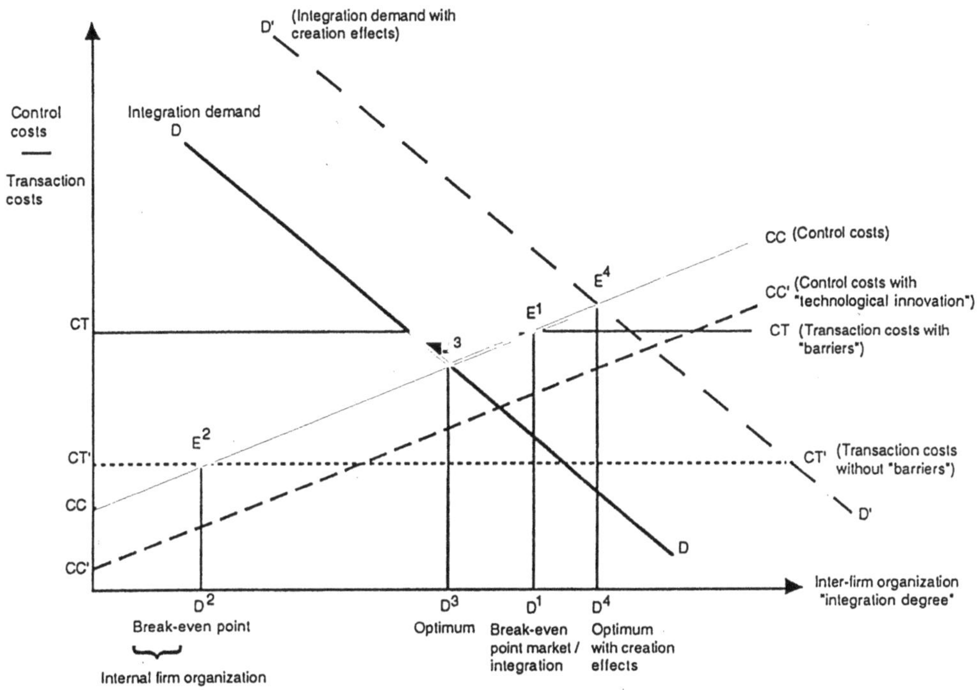


Source: R. Ratti (1990)

**Fig. 6.1** The degrees of integration of a firm (transaction costs, control costs and preference to the integration)

- The reasoning has been conducted in terms of minimized costs. To the objection that the resilient integration requires also a creation effect, it is possible to answer, in this case, that the preference curve can be displaced upward (from D-D toward D'-D'). It will also determine an optimum (E4) which will even be above the transaction costs, having as a consequence a particularly high integration degree.

Finally, even supposing that the trend of some transaction costs is leaning downward, following the Europe without borders postulates, the relevance and the continuity of an organization in a network depends on its capacity to innovate, both in terms of minimizing control costs and in terms of value creation.

In conclusion, in the new context of an open border, the theoretical analysis advocates that the economic development of the border areas will not be determined by the political-institutional differential and therefore by the differentiated position of profits, positive and negative, due to the effect of belonging to one or the other nation, but more likely, by the compared real advantages of both border areas. The 'open border' implies the passage from the concept of border areas economy to that of transborder economy. This situation can imply some quick and fundamental adjustments. The strategic behaviour of the partners is particularly crucial.

A strategy in terms of a cooperation network is the most efficient approach to overcome persistent or residual obstacles and uncertainty situations typical of a border context. But, all this needs to be supported by a strategy of functional synergies, capable of realizing themselves at a level covering the whole transborder area in a timely manner as in the case of the filter-border.

## 6.3 The Overcoming of the Barrier/Border-Obstacle by the Functional Space Analysis: The Supporting Space

In addition to the microeconomic point of view, analyzed through the strategic behaviour of the parties, there is a second theoretical possibility to interpret the border particularly in its function as a contact area, represented in terms of spatial analysis i.e. the study of the strategic spaces of enterprises and, particularly, what has been called the supporting space, that is the ensemble of the factor-frames and of the relations 'preceding the market'.

### 6.3.1 Postulates

The 'strategic' or 'life' spaces of an enterprise are determined by three functional spaces (Ratti 1971; Ratti and D'Ambrogio 1988; Ratti, 1989):

- The 'production space' of the firm is determined by the spatial division of the work following the model of the segmentation theory. An enterprise buying outside will define and delocalize its production, following the technological, economics and socio-cultural characteristics specific to each segment and to each production region.
- The 'market space' is determined by the relation that the enterprise entertains with its different markets. These spatial relations are characterized by the number, intensity, structural characteristics and by the evolution process of these markets, relative to their environment.

This definition, in spatial-functional terms of the offer and demand, already constitutes a significant progress in the dynamic approach of the enterprise development. However, it still seems insufficient to deal with the interior and exterior strategic aspects of the enterprise. It is therefore proposed to define a third functional space of the enterprise.

- The 'supporting space' which describes three types of pre-competitive relations:
- The qualified or privileged relations at the organization level of the production factors (capital origin, information source, technological ability, particular ties at the human capital level, etc.);

- The strategic relations of the enterprise with its partners, suppliers or clients (privileged informations exchange, cooperation, partnership, alliance, partial integration, etc.);
- The strategic relations with the territorial environment parties (public institutions, private or semi-public associations, etc.).

It is possible to represent in Figure 6.2 the characteristic traits of the spatial-functional relations of an enterprise following the traditional non-spatial model. Following this model the strategic space of the enterprise is then defined.

### 6.3.2 Thesis

- Among the three functional spaces of the enterprise, the supporting space appears to be particularly crucial to the strategic orientation of the enterprise. Specifically it allows to directly take into consideration the so-called factor frames and also some important relations of pre-competition character.
- The aggregation of functional spaces of the different enterprises can, with a certain possibility (Christensen 1988), give way to some spatial functional relations. Territorially, in terms of polarized space or of functional space 'force fields', the supporting space dynamics push the enterprises to look for common interests and to cooperate with the territorial institutions for the creation of a favourable environment.
- This relative coincidence between a strong node of functional spaces and territorial spaces allows 'local synergies'. This is defined as the result of a territorial materialization of an ensemble of functional relations structured in terms of a pole or force field. This logic leads to other notions applied in the recent regional literature: 'local industrial network', 'localized ecosystem', 'industrial district', 'localized industrial system' and, finally, towards the notion of 'milieu' developed by the GREMI group (1985, 1988).

### 6.3.3 Corollaries

- The two border characteristics (barrier or contact) influence in a determinant manner the strategy spaces of the enterprises, particularly their supporting spaces.
- Where the separation factors are dominant, the possibility that the border area is the origin of the creation of a real supporting space is insignificant. Often, the existence of this type of space is strongly dependent on the politics of the States-Nations and entry port. The eventual synergy effects have a temporary and unilateral character, with the exclusion of 'regional' effects.
- Where the contact factors (effect of economic, social and cultural proximity) are dominant, the border can constitute a potentially favourable element for the determination of a supporting space. The probability for synergies to appear at a regional level will be stronger, and essentially dependent on the strategic determination of the regional parties. History shows that each town or port

where the localization is favourable to exchanges, has developed a real 'supporting' strategy (custom tax free towns, duty free fairs, etc.). A recent and very pertinent example within the framework of the Europe of Regions context is the Regio Basiliensis, created about thirty years ago.

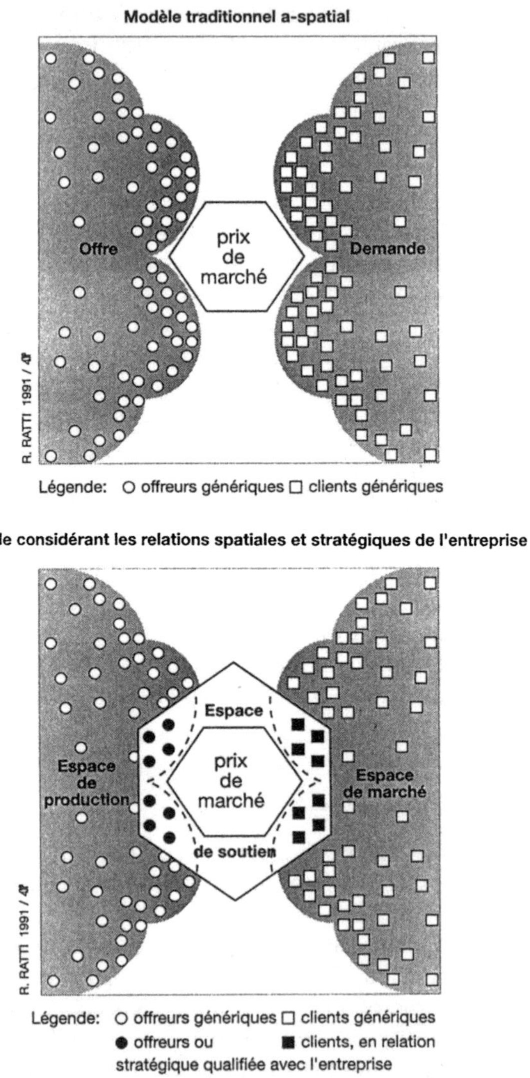


**Fig. 6.2.** The economic-functional strategic spaces of a firm

## 6.4 Conclusions

It becomes evident with the economic and political necessity tied to the creation of a real space of a free European market that the 'open border' implies the passage from the concept of a border areas economy to that of transborder economy. This situation can imply some quick and fundamental adjustments. The strategic behaviour of the partners is particularly crucial. A strategy in terms of a cooperation network is the most efficient approach to overcoming obstacles and uncertainty situations typical of a border context. However, all this needs to be supported by a strategy of functional synergies, capable this time of realizing themselves at a level covering the whole transborder area in a timely manner as in the case of the filter-border.

Therefore, the border in its contact function needs to be pursued positively as a pre-condition to the opening of space.

## References

Biucchi BM, Gaudard G (1981) Régions frontalières. Ed Georgi, St. Saphorin, Switzerland

Button K, Rossera F (1990) Barriers to communication. A literature review, The Annals of Regional Science 24:337-357

Camagni R (Ed) (1991) Innovation Networks. Belhaven Press, London and New York

Christensen PR (1988) Enterprise Flexibility and Regional Networks. Paper presented at RSA Annual Congress

Coase RH (1937) The nature of the firm, Economica 4

Covin JG, Slevin DP (1989) Strategic Management of Small Firms in hostile and benign Environments. Strategic Management Journal 10:75-87

Di Tella G (1982) The economics of the frontier. In: CP Kindleberger and G Di Tella, Economic in the long view 1. Models and methodology. Macmillan, London and Basingstoke

Faludi A (1986) Toward a Theory of Strategic Planning. Netherland Journal of Housing and Environmental Resources 3:253-268

Guichonnet P, Raffestin C (1974) Géographie des frontières. Presses Univ. de France, Paris, France

Hansen N (1977) Border Regions. A critique of spatial theory and a european case study. Annals of Regional Science 11:1-14

House JW (1980) The frontier zone. A conceptual Problem for Policy Makers. International Regional Science Review 1:456-477

Kamann D, Strijker D (1991) The network approach. Concepts and applications. In: R Camagni (Ed) Innovation Networks. Spatial Perspective. Belhaven Press, London and New York

Loesch A (1940) Die räumliche Ordnung der Wirtschaft. G Fischer, Jena

Nijkamp P, Rietveld P, Salomon I (1990) Barriers in Spatial Interactions and Communications. A Conceptual Exploration, Annals of Regional Science, 24:237-252

Porter ME (1980) Competitive strategy. Techniques for analyzing industries and competitors. Free Press, New York

Porter ME (Ed) (1986) Competition in global industries. Harvard Business School Press, Boston, USA

Ratti R (1971) I traffici internazionali di transito e la regione di Chiasso. Editions Universitaires, Fribourg, Switzerland

Ratti R (1988) Development Theory, Technological Change and Europe's Frontier Region. In: Ph Aydalot, D Keeble, High Technology Industry and Innovative Environments. The European Experience. Routledge, London and New York

Ratti R (1990) The Study of the spatial effects of the borders. An overview of differents approaches. NETCOM 4:37-50

Ratti R (1992) Innovation technologique et développement régional. IRE/Méta éditions, Bellinzona, Switzerland

Ratti R, Reichman S (Eds) (1993) Theory and Practice of Transborder Cooperation. Helbing & Lichtenhahn, Basel und Frankfurt am Main

Reichman S (1989) Barriers and Strategic Planning - A Tentative Research Formulation. Paper presented at NECTAR Meeting, Zürich, Switzerland

Richardson GA (1972) The organisation of industry, The Economic Journal 82:883-896

Turner FJ (1921) The Frontier in American History. New York, USA

Williamson OE (1975) Markets and Hierarchies. Analysis and Antritrust Implications. Free Press, New York; Collie Macmillan Pub, London

Williamson OE (1985) Economic Institutions of Capitalism. Firm, market, Regional Contracting. Free Press, New York; Collie Macmillan Pub, London

Williamson, OE (1986) Economic Organization. Firms, Market and Policy Control. Wheatsheaf Books, Brighton, UK

# 7 International Migration in Europe: Overcoming Isolation and Distance Friction

P. Nijkamp and K. Spiess

## 7.1 Introduction

In discussing the theme of 'overcoming isolation' we are usually inclined to look at physical barriers preventing spatial interaction. Missing links in networks, infrastructure bottlenecks or geographical peripherality are the obvious examples of impediments to free movement in space. In such a context, interregional or international trade flows, commuting, congestion, accessibility and network performance are normally discussed. Far less attention has been given in Europe to the issue of international migration, either labour migration or forced migration as a result of geo-political developments in Southern, Central and East-Europe. In recent years we witness a concern on massive migration flows into the West-European space, a phenomenon which reflects the fact that Europe is still socially fragmented and only politically more open (see Nijkamp and Spiess 1993). Therefore, it makes sense in a publication on overcoming isolation to pay explicit attention to facts and backgrounds of international migration in Europe. The completion of the internal market by 1993 provoked much debate on the consequences of a free mobility of goods, people and information in the EC countries. Also the expected migration waves from former communist countries in Central and East Europe created an intensified concern on the EC as a magnet for international migrants (cf. Ghosh 1991). And finally, the increasingly important phenomenon of illegal migrants in Europe led to doomsday scenarios of the U.S. - Mexico border type. It seems as though Europe is now entering the 'age of migration' (see Castles and Miller 1993).

It should be noted that even without the above recent megatrends the EC countries in Europe - with the clear exception of Ireland - have in the past decade shown a significant rise in the share of foreign people. Given the current aging trends, the EC countries - again with the exception of Ireland - would in the future even lose population without immigration (see also Nijkamp et al 1991).

The debate on the extension of the European space provokes thus many policy issues on migration. The question whether economic integration in Europe will generate significantly higher migration flows is not very easy to answer, but a somewhat older study by Böhning (1974) shows that in the past the demand for labour force for most European countries was a far more important motive than integration per se. In this context, Feithen (1986) has more recently demonstrated that for international labour migration in European push-pull factors (e.g. high unemployment rates in the country of origin) appear to be of decisive importance. Observations from recent years confirm that massive migration flows as a result of economic integration in Europe have not taken place, and as far as migration has taken place there is not a significant difference between less and more developed countries (cf. Garson 1992).

Apart from a concern on intra-EC migration, there is nowadays an increasing concern about migration into EC countries from so-called third countries. The past years have shown an increasing flow of third country in-migrants, subdivided into asylum seekers, Aussiedlers (ethnic Germans) and remaining migrants. Apparently, the EC countries have become less isolated from the remaining part of Europe and other parts of the world (see Muus and Cruijsen 1991).

In view of the rising flows of third countries in-migrants, all EC countries have adopted in the meantime a more restrictive in-migration policy. Consequently, the foreseeable future in-migration flows in EC countries will depend on various critical factors:

- the socio-economic conditions in the countries of origin
- the socio-economic perspectives (and needs) in the country of destination
- the socio-cultural distance between countries of origin and destination (e.g., the existence in the country of destination of a critical mass of foreign people with the same cultural background or the existence of a favourable absorption mechanism for foreigners)
- the existence of regulatory policy regimes for foreign in-migrants (including enforcement of regulations against illegal migration).

The relative importance of the above factors is largely unknown as yet. International migration has therefore become an important issue in social science research, but it is a field fraught with many uncertainties. In light of the above observations, the present paper serves to address critical factors (including policy regulations) that impact on international migration in the European context, with particular emphasis on empirical information.

After the presentation of some trends and facts in Section 7.2, we will present in Section 7.3 a concise survey of relevant explanatory frameworks for international migration. This will be followed by a final section on absorption factors and policy factors regarding migration in receiving countries.

## 7.2 Past Trends

A glance at European migration in the past two centuries shows that by and large Europe may be characterized as an emigration continent. Before the Second World War it was not Europe that absorbed foreigners, on the contrary it was predominantly the Europeans who were mainly absorbed by the three big immigration countries, the United States of America, Canada and Australia (e.g. Borrie 1990). Talking about Europe - and especially Western Europe - as an immigration continent was certainly not possible before the Second World War. And even in the post-war period not all Western European countries have been immigration countries for this whole period, as is shown by Table 7.1 for the period 1960-1988.

For the year 1990, a comparison between selected European countries (see Table 7.2) shows that only Ireland - with negative net migration - may nowadays be characterized as an emigration country.

We will now in a concise manner depict the different types of international migration flows in Europe since the Second World War (see Borrie 1992; Coleman 1993; Fassmann and Münz 1992; Zlotnik and Hovy 1990). Next, we will describe the present composition of foreign population in six selected countries of the European Community and explore whether it is possible to identify special patterns (see Fassmann and Münz 1992). Finally, we will take a short look at the distribution of various types of migration over selected European countries (see Coleman 1993).

A first type of migration flow into Europe, which has taken place in different time periods since 1950, is linked to the postwar *decolonization process*. Besides the return of European colonists and colonial officers, countries like Great Britain, France, Belgium, the Netherlands and Portugal, which had colonies in the past, recorded sizeable immigration flows by migrant workers from their former overseas territories.

A second type of immigration flows played a major role in the 1950s until the mid 1970s, when a number of countries - notably Switzerland, France, the Federal Republic of Germany, the Benelux countries and Sweden - started to meet part of their growing demand for labour by recruiting a labour force in several Mediterranean countries. In the initial period of hiring these *guestworkers*, Italy was the most important recruitment area, while in the 1960s Spain and Portugal became the most important recruitment and emigration countries. In the 1970s, Turkey, Morocco and Tunisia became increasingly supply countries of migrant workers. In the early 1970s, the employment of foreign labour reached a peak in the post-war period. In 1970, West Germany was leading with nearly 2.1 million foreign workers.

After the oil price shock and the economic recession in the mid 1970s the recruitment of foreign labour drastically declined and in many countries restrictive immigration regulations were imposed. In view of the reduced absorption capacity of the labour market, the aim was to stop further immigration. But these policies had only a short term effect. This is especially true for Germany, where the average annual net migration flows of foreigners

**Table 7.1.** Average annual migration flows (citizens and foreigners) to and from selected European countries, 1960-1988

| | Period | | | | | |
|---|---|---|---|---|---|---|
| | 1960-1964 | 1965-1969 | 1970-1974 | 1975-1979 | 1980-1984 | 1985-1988 |
| Immigrants | | | | | | |
| Belgium | 69,056 | 65,583 | 64,688 | 58,271 | 47,862 | 48,584 |
| Germany, Fed. Rep. | 576,211 | 706,144 | 873,051 | 527,483 | 502,179 | 554,230 |
| Nether-lands | 57,746 | 71,009 | 89,140 | 97,571 | 79,419 | 88,481 |
| Sweden | 29,226 | 45,404 | 43,342 | 41,368 | 32,212 | 41,593 |
| United Kingdom | - | - | - | 186,600 | 186,400 | 227,500 |
| Emigrants | | | | | | |
| Belgium | 35,271 | 41,503 | 47,615 | 52,990 | 58,663 | 54,949 |
| Germany, Fed. Rep. | 364,108 | 508,695 | 566,839 | 521,131 | 499,138 | 411,128 |
| Nether-lands | 51,218 | 60,329 | 61,179 | 59,737 | 61,962 | 54,434 |
| Sweden | 15,226 | 19,842 | 35,697 | 23,897 | 27,396 | 22,166 |
| United Kingdom | - | - | - | 207,600 | 214,000 | 208,500 |
| Net Migration | | | | | | |
| Belgium | 33,785 | 24,080 | 17,073 | 5,281 | -10,801 | -6,365 |
| Germany Fed. Rep. | 212,104 | 197,449 | 306,211 | 6,352 | 3,040 | 143,102 |
| Nether-lands | 6,528 | 10,679 | 27,961 | 37,833 | 17,457 | 34,047 |
| Sweden | 14,000 | 25,562 | 7,644 | 17,472 | 4,816 | 19,427 |
| United Kingdom | - | - | - | -21,000 | -27,600 | 19,000 |

Data for the Federal republic of Germany cover only 1961-1987. Those for Belgium refer to 1960 and 1963-1987. For the United Kingdom, the period covered is 1975-1988.
Source: Zlotnik and Hovy (1990, Table 1)

**Table 7.2.** Recent gross inflows of migrants into selected European countrieS, 1990, by type (thousands)

| Country of destination | Labour migrants | Ethnic migrants | Asylum seekers (gross inflow) | All foreign (excluding asylum seekers, includes some EC) | Total gross inflow, including asylum seekers | Total population | Gross immigration per 1,000 (includes asylum seekers) | Net migration per 1,000 |
|---|---|---|---|---|---|---|---|---|
| Austria | 103.4 | | 22.8 | 123.1 | 145.9 | 7,660 | 19.0 | 16.1 |
| Belgium | 3.7 | | 13.0 | 52.3 | 65.3 | 9,948 | 6.6 | 1.1 |
| Denmark | | | 5.3 | 40.7 | 46.0 | 5,135 | 9.0 | 1.7 |
| France | 26.2 | | 54.7 | 117.8 | 117.8 | 56,304 | 2.1 | 0.9 |
| Germany | 138.6 | 397.1 | 193.1 | 649.5 | 1239.7 | 62,679 | 19.8 | 15.9 |
| Greece | | | 4.1 | | | 10,057 | | 2.9 |
| Ireland | | | | | | 3,506 | | -2.6 |
| Italy | | | 4.7 | | | 57,576 | | 2.4 |
| Netherlands | | | 21.2 | 60.1 | 81.3 | 14,893 | 5.5 | 3.3 |
| Norway | | | 4.0 | | 15.7 | 4,233 | 3.7 | 0.4 |
| Portugal | | | 0.1 | | | 10,337 | | 0.9 |
| Spain | | | 8.6 | | 42.5 | 38,925 | 1.1 | -0.2 |
| Sweden | | | 29.4 | | 53.3 | 8,527 | 6.3 | 4.1 |
| Switzerland | 46.7 | | 35.8 | | 137.2 | 6,674 | 20.6 | 8.5 |
| United Kingdom | 34.6 | | 30.0 | | 82.4 | 57,323 | 1.4 | 0.9 |
| Total | 353.2 | 397.1 | 426.8 | | 2027.1 | 353,778 | 5.7 | |

Notes: Blanks indicate that data are not available. Data on labour migrants to EC countries do not usually include citizens of other EC countries

Austria: 'All foreign' figure is an estimate of net migration.

France: all foreign excludes some EC. Belgium: labour data 1989. Germany: all foreign 1989

Germany: data refer to the area of the Federal Republic befgore re-unification. The total of ethnic migrants ferers only to 'Aussiedler' from Eastern Europe and the USSR, not from the former East-Germnany, of whom there were 238,282 from January 1990 to June 1990.

All foreign figure is for 1989. 'Total inflow' for 1990 includes that figure (Meyer 1992)

Spain: 'All foreign' total is for 1989. Switzerland: excludes seasonal workers.

Net migration includes persons of all citizenship. Foreign immigrants excludes asylum seekers.

Source: Coleman (1993, Table 1)

decreased from 297,040 in the period 1970-74 to -32,370 in the period of 1975-79 (Zlotnik and Hovy 1990). The long-term effect however, shows no reduction in immigration flows: after 1985 almost all Western European countries experienced a recovery of their annual immigration flows of foreigners.

Despite the formal cessation of labour migration, it is noteworthy that in the meantime this situation has led to a rise in migration for family reunification. This third type of European post-war migration, namely *family migration*, makes up one of the major migration flows in the recent past and present time. Furthermore, a new phenomenon has emerged in this context, viz. family formation migration, that means a marriage with a foreign partner. Finally, an important side-effect of the increasing restrictive immigration regulations since the mid 1970s should be added, viz. the rising number of *illegal immigrants* into Western Europe.

Since the 1980s a new pattern of labour migration developed. Besides the North-West European countries *also* most *Southern European countries*, particularly Italy, *have become countries of immigration*. In addition it is noteworthy that, much of the 'official' labour migration within and into Europe at present is "high level manpower [...] and to a lesser extent, with high level manual skills" (Coleman 1993 p.14).

With growing political conflicts, civil wars and economic crises in the Middle-East, South America and Africa in the recent past - in addition to the above mentioned increasing flows of foreign labour and family migration -, the flow of *asylum seekers* into Western Europe is increasing since the 1980s. In 1991, Western Europe received 539,000 applications for asylum (Coleman 1993). As a special case, the emigration flows in 1991/1992 out of the former Yugoslavia - mainly as a result of the war between Croatia and Bosnia-Herzogovina - has to be mentioned, as this is the largest single wave of emigration since 1946/47 (Coleman 1993).

The above mentioned types of migration may mainly be characterized as *south-north* migration. *East-west* migration within, in and out of Western Europe has almost never played a significant role in the post-war European migration history. East-west migration was largely suppressed for 40 years due to the political division of Europe. Mass migration has only occurred in cases of political crisis. With the end of the socialist regimes in Central and Eastern Europe however, east-west migration flows have been intensified. Particularly the large flows of people with a German origin out of Poland, the former Soviet Union, and Rumania should be mentioned in this context.

Besides these main patterns of European postwar migration, Fassmann and Münz (1992) mention also another pattern in European migration: the migration between *hinterlands and home countries*. This means that European migration clearly shows privileged social and geographical relationships between different countries. A further demonstration of this statement can be found in an analysis of the share of foreign resident population in six major receiving countries in Western Europe.

As Table 7.3 indicates, almost all Algerians, Tunisians, Portuguese and Moroccans in these six countries live in France. The vast majority of migrants from East-Central Europe is to be found in Germany, while in addition, some 72 per cent of all ex-Yugoslavs and 74 per cent of Turks reside in Germany. Nine of the ten Greeks living in the six major receiving countries also reside in Germany. Most Finnish emigrants moved to Sweden, while most Austrian migrants emigrated to Germany. Although not in Table 7.3, it is well-known that the United Kingdom absorbs almost all Irish (outside Ireland), and almost all Indians, Pakistanis and Bangladeshis living in Europe. Migrants from Italy and Spain follow a less specific pattern. Italians outside Italy, who are registered in the six major receiving countries are mainly spread over Germany (37.8 per cent), Switzerland (26.7 per cent) and France (17.5 per cent). The majority of Spanish foreigners can be found in France (44.4 per cent), Germany or Switzerland.

From a total of 13.4 million foreigners in the EC in 1989, 49 per cent were citizens of non-European countries, 38 per cent citizens of other EC states, while only 13 per cent were citizens from other European countries. Among the non-European countries the Turkish people are, with 31 per cent of all Non-European citizens in Europe, by far the largest group, while all Africans represent about 36 per cent. Migrants from Asia and South and North America together represent about 33 per cent of the foreign citizens of non-European countries (Coleman 1993).

Finally we will focus upon the question, whether special types of migrants can be associated with special countries. A look at Table 7.2 containing data on the gross flows of migrants into Western Europe by types of immigration (labour migrants, ethnic migrants and asylum seekers) shows that ethnic migrants are only to be found in Germany. Muus and Cruijsen (1991) also mention in this context the ethnic Greeks coming from Eastern Europe. For the group of asylum seekers, it is noteworthy that Germany - followed by France - seems to be the most attractive country for them. The majority of labour migrants also seems to prefer Germany as the most attractive country of destination. Austria appears to be positioned on the second place in terms of absolute numbers of labour migrants. These high (absolute) numbers of all types of migrant inflows to Germany is parallelled by the fact that this country has in absolute numbers the largest foreign population, even without taking the ethnic Germans into account. In a relative sense however, by taking the foreign population as a percentage of the total population, Switzerland is with 16.5 per cent foreign population, the European country with the highest percentage of foreigners among its population (Coleman 1993).

## 7.3 Analytical Frameworks for Explaining International Migration: A Concise Survey

There is not a single theory for the explanation of international migration. There are different causes for international migration, there are different views on backgrounds and motives of migrants and hence there are different explanatory frameworks. We will only give a concise overview here.

### 7.3.1 The Equilibrium Approach

**Table 7.3.** Foreign resident population in six major receiving countries of Western Europe: Per cent distribution of foreign residents by country of origin within the total foreign population of that origin residing in six receiving countries, and total foreign resident population by country of origin and country of residence, 1990

| | Country of residence | | | | | | Six-country total | |
|---|---|---|---|---|---|---|---|---|
| Country of origin | Belgium | France | West Germany | Nether -lands | Sweden | Switzer-land | Percent | FRP (1,000s) |
| Austria | * | * | 85.2 | * | 1.3 | 13.5 | 100.0 | 213 |
| Finland | * | * | 7.9 | * | 92.1 | * | 100.0 | 130 |
| Greece | 5.9 | * | 88.6 | 1.4 | 1.8 | 2.3 | 100.0 | 355 |
| Italy | 16.6 | 17.5 | 37.8 | 1.2 | 0.3 | 26.7 | 100.0 | 1,452 |
| Portugal | 2.0 | 76.8 | 10.1 | 1.0 | * | 10.2 | 100.0 | 841 |
| Spain | 9.7 | 44.4 | 27.6 | 3.5 | 0.6 | 23.9 | 100.0 | 539 |
| Turkey | 3.8 | 8.9 | 74.3 | 9.0 | 1.1 | 2.8 | 100.0 | 2,255 |
| Yugo-slavia | 0.6 | 5.7 | 72.1 | 1.5 | 4.5 | 15.5 | 100.0 | 905 |
| Algeria | 1.7 | 97.3 | 1.1 | * | * | * | 100.0 | 637 |
| Morocco | 14.9 | 61.5 | 7.1 | 16.5 | * | * | 100.0 | 951 |
| East-Central Europe | * | 11.0 | 81.0 | * | 5.7 | 2.3 | 100.0 | 421 |
| Tunisia | 2.6 | 85.6 | 10.7 | 1.1 | * | * | 100.0 | 242 |
| United States | 28.7 | * | * | 27.9 | 19.6 | 23.8 | 100.0 | 41 |
| Others | 10.2 | 25.8 | 39.6 | 8.5 | 8.2 | 8.5 | 100.0 | 3,050 |
| Total (percent) | 6.9 | 30.2 | 43.8 | 5.8 | 4.0 | 9.3 | 100.0 | |
| Total (1,000s) | 905 | 3,608 | 5,242 | 692 | 484 | 1,100 | | 12,030 |

Note: Asterix indicates no data are available.
Sources from Fassmann and Münz: OECD.SOPEMI (1992, several tables)
Source: Fassmann and Münz (1992, Table 3)

The *equilibrium* approach to international migration represents a main direction in the explanation of international movements of people (see for an evaluation Wood 1982 and Bach and Schraml 1982). The equilibrium model of migration focuses the attention primarily on the rational calculus of the individual action.

Migration is in general regarded to be an individual choice process which can be cast in the framework of a neoclassical explanatory model. Migration streams are seen as the cumulative result of individual decisions based on a rational evaluation of the benefits to be gained and the costs entailed in moving. In the long run, migration will lead to a gradual convergence in the level of economic growth and social well being. In the context of labour market absorption it should be noted that the equilibrium approach seeks critical success absorption factors mainly in employment opportunities with higher returns.

### 7.3.2 The Historical Structural Approach

A second main direction of migration theories, the *historical-structural approach*, focuses on the origin of the costs and benefits faced by the potential migrants (see for a review again Bach and Schraml 1982; Wood 1982). Migration is seen as a macro-social process. Even though this approach is - because of the variety of considerations included - not so easy to characterize as the above equilibrium approach, in general this approach considers migration from a broader context of socio-economic and political changes. The bases of the structural perspective are mainly to be found in the historical materialism, according to which migration is deeply rooted in the pressures and counter-pressures in national economies, which lead to changes in the organization of production. Structural factors influence labour mobility through their impact on the degree and the spatial distribution of the demand for labour. Patterns of migration are thus explained in terms of changes in the organization of production which unequally affect the fortune of different classes. Critical success absorption factors are in general structural factors like socio-economic and political developments of capitalist economies.

### 7.3.3 The Utility Maximizing Approach

The utility framework offers a more modern version and extension of the economic equilibrium approach to migration (see, for instance, Borjas 1989). The author points out that this more recent approach to the economics of (im)migration is based on main assumptions of neoclassical theory. This means that the individual is a utility maximizer, whose behaviour is constrained by the financial situation of the actor and by the migration policies of the source country and the immigration policies of potential host countries. Based on the assumption that exchanges among various players lead to an equilibrium in the market place, Borjas discusses extensively the existence of an immigration market, which is acting as a sieve for migrants across potential host countries. In this immigration market, the different host countries function as the suppliers, making migration 'offers' in respect to a certain set of immigration regulations from which individuals on the demand side compare and choose, and so allocate themselves in the end nonrandomly among countries. Especially interesting in our context is his assumption that host countries with a certain set of immigration regulations attract different types of persons. Based on these main assumptions, Borjas concentrates then on three questions, the determinants of the size and skill composition of immigrant flows to any particular host country, the process of assimilation of migrants in the host country, and the adjustment process after immigrating in the host countries' labour market.

### 7.3.4 The Welfare State Approach

An interesting class of migration analyses refers to the importance of the welfare state in connection with international migration (cf. Freeman 1986). Freeman's approach, which may be characterized as a welfare state approach, may be positioned in the field of political economy. The starting point of his analysis is to understand the development of a welfare state as a dialectic phenomenon between the distributive logic of closure and distributive logic of openness. The former is related to the fact that the welfare state has to restrain entry to preserve the advantages of itself; it has to restrict benefits and rights only to members. But at the same time the welfare states are on the other hand deeply embedded in the global political and economic order, and it is to this fact that the latter logic refers. The national economies of the welfare states are engaged in systematic exchanges with this larger global system.

In this openness however, Freeman sees several problems for the welfare state as well. Looking at the wide inequalities between the benefit levels and living standards that exist both among welfare states themselves and between them and the outside world, he notes that these inequalities are the main reason for external pressure from the international economy to disrupt and threaten the privileges that the welfare state represent. As the most important and directly relevant external economic factor from the point of view of the welfare state, Freeman addresses the issues of foreign labour. In line with the classical economic migration approach, Freeman seeks the explanation for migration in wage differences. But in contrast to the classical economic approach, he underlines the importance of indirect wages in the context of international migration. "Along with the high real, direct wages the social wage is part of the package of compensation that exerts an attractive pull on workers in less prosperous societies drawing them to the rich countries in anticipation of better lives" (Freeman 1986, p.55). But for Freeman it is not just a simple attraction of migrants by the welfare state, but rather the availability of the welfare state benefits to indigenous workers which helps the sequence of events which create the demand for foreign labour. In this context foreign labour is seen as the only real alternative to the elimination of the privileges of indigenous work force, but only if new workers are excluded from the rights of the welfare state. In this inflexibility of the welfare state's labour market, Freeman looks amongst others for an explanation of the guest worker systems.

### 7.3.5 The Regulatory Approach

Migration is not just a free movement of people, but is strongly influenced by various physical and non-physical barriers. Zolberg (1989) has drawn attention to the importance of borders and regulations for migration. His approach aims to deal with the question that migration theories in general ignore the political dimension of migration. According to him it is the political perspective which helps to overcome the limitations of the historical-structural and the economic equilibrium approach. This political approach should - in this view - be macro-analytical and historical in nature. His framework consists of a world of individuals who maximize their welfare by exercising a variety of choices from which migration is only one. A second element of this world are exclusive societies acting as organized states to maximize collective goals by controlling the exit or entry of individuals. In Zolberg's approach it is very important that the character of international migration is related to a fundamental tension between the interests of individuals and societies and in addition between the sending and receiving countries.

Taking into account the fact that several states interact as parts of a larger whole - a situation which Zolberg calls an international social system - he takes the overall structural configuration of the international social system to provide and approximate an analytical matrix for analyzing migration policies. He illustrates his framework by two topics, labour migration from the Third World and refugee flows from new states.

Concerning the first topic the author studies the conflicting interests of industrial societies to maximize labour supply and to protect cultural integrity. He comes to the conclusion that, wherever economic expansion grinds to a halt, other concerns - like the integrative - get a greater impact on policy-making. The tension of migration may be found in the fact that it is quite evident for the population of less favoured regions to migrate to more desirable countries, but on the other for the more fortunate countries it is rational to preserve their favourable endowment by restricting entry. Only the self interest of the more favourable regions to obtain labour can in the end lead to entry possibilities.

### 7.3.6 The Tension Approach

There are only a few approaches to international migration which try to explain migration in the context of one general, global theory. The approach of Hoffmann-Nowotny (1981) is an example of such an attempt, which we may characterize - according to Penninx (1986) - in comparison to other more inductionistic-oriented attempts a deductionist approach. Hoffmann-Nowotny (1981) aims to explain international migration as a specific field of social reality only by means of the general Theory of Societal Systems. His System Theory Approach allows for a discussion of migration on the level of three different system units: the individual, the class and the collective system unit. Characteristic for Hoffmann-Nowotny's analysis on all levels is that migration is seen as the result of structural and anomic tensions. Structural tensions are seen as the result of a divergence of power and prestige, which are themselves regarded as central theoretical concepts corresponding to the dimensions of societal systems conceptualized via structure and culture. Anomic tension is seen as the empirical consequence occurring at the moment when structural tensions exceed a certain threshold level, which is not further specified. From in this theoretical context, Hoffmann-Nowotny is analyzing migration as a process by which tensions are transformed and transferred.

### 7.3.7 The System Approach

The approach which has been discussed most in the most recent past and has increasingly gained attention is the system approach. This approach is based on the recognition that to capture the changing trends and patterns of contemporary international migration (see Section 7.2) requires a dynamic- instead of a static-perspective. Besides this emphasis on dynamics, a further main characteristic of the system approach is that a so-called migration system is used as the basic system of analysis (Kulu-Glasgow 1992). A migration system is defined as two or more places or more specific countries connected to each other by flows and counterflows of people.

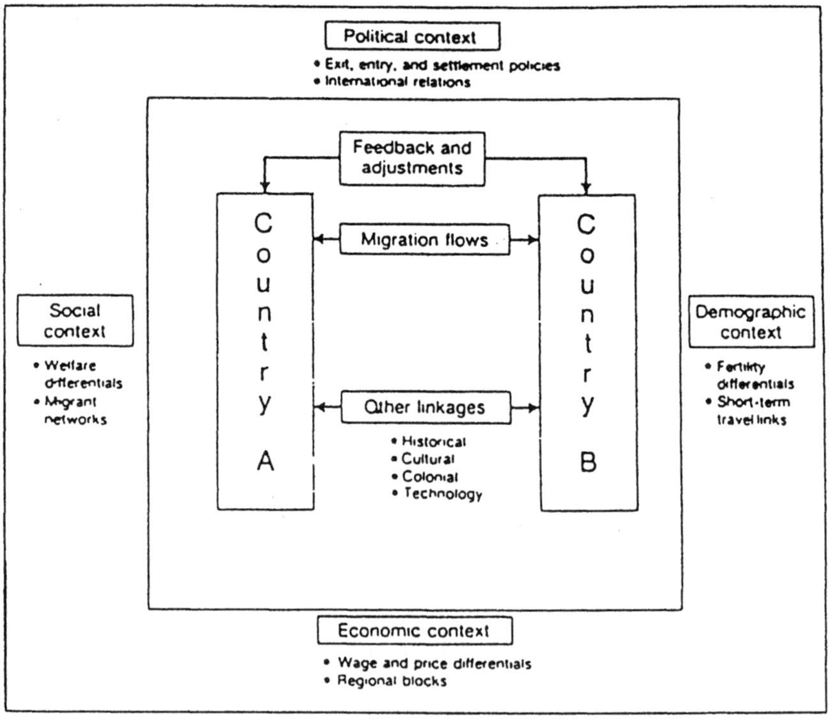


Source: Kritz and Zlotnik (1992, p 3)

**Fig. 7.1.** A system framework of international migration

Figure 7.1 presents a scheme of a system approach to international migration by Kritz and Zlotnik (1992) indicating that besides flows of people also other flows link countries together in a system. Such flows occur within national contexts whose political, demographic, economic and social dimensions are changing partly in response to the feedbacks and adjustments that stem from the migration flow itself. In general, receiving countries are characterized by higher wages and better welfare, while sending countries with high numbers of emigrants have as - Kritz and Zlotnik (1992) note - usually lower wages and poorer conditions. If international migration is analyzed as embedded in a system like the one described in Figure 7.1 it is evident that micro- and macro-elements are both involved in the analyses. The role which the individual has in this system is the role of an active decision-maker, who develops strategies to migrate which are embedded in the different influences of the system.

At the end of this concise overview we may conclude that a system framework has - compared with the other approaches - the big advantage that it tries to take into account the variety of factors which play a role in the migration process. This approach is not restricted to a special type of migration and it does not only explain the existence of migration but also their size and composition over time. The special emphasis on networks - and especially on social networks - has a far greater importance if we recognize the fact that nowadays a majority of migration flows may be characterized as family migration. But on the other hand, also the system approach like most of the above approaches has only a framework character, which in the end does not allow to specify functional relationships or to offer directly testable hypotheses in empirical research.

## 7.4 Absorption Factors in International Migration

In this final section we will address the issue of the necessary conditions for a succesful migration; in other words, which attraction and regulatory forces play a critical role in choosing a specific country of destination?

At the outset it should be noticed that the current scientific knowledge on critical success absorption factors is so far not very well developed. In most cases various aspects of critical success absorption factors in migration are analyzed as only one element next to many others, while empirical studies are not abundant in number. Important contributions can be found in studies by Engelbrektsson (1982) and Boisvert (1987), who focus on the role of social networks in the migration process, respectively regarding Turkish migrants to Sweden and Portuguese migrants to France.

For a meaningful classification of critical success absorption factors we refer here to an article by Fawcett and Arnold (1987). In their approach - which belongs to the broader group of system approaches - they present a migration system operating in a set of structural and contextual factors. The authors point out three different groups of such structural and contextual factors, namely

7.4.1 *state to state relations and comparisons*, 7.4.2 *mass culture* connections and 7.4.3 *family and social networks*. These structural and contextual factors may according to Fawcett and Arnold (1987) explain the diversity of Asian and Pacific immigrants to the United States. These three broad classes of contextual factors will now briefly be discussed.

### 7.4.1 State to State Relations and Comparison

Since in general economic aspects play a major role in explaining migration, it seems plausible to look first at a *higher level of economic development* as an indicator related to 'state to state relation and comparison'. The characterization of this factor as an critical success absorption factor is underlined by the observation that all major receiving countries have higher developed capitalistic economies and at least the image of an open economic system. As Fawcett and Arnold (1987) point out, the difference in economic development between origin and destination may be even more pertinent than the structure of the economic system.

Not mentioned by Fawcett and Arnold (1987), but of great importance from the angle of the welfare state approach is another critical success absorption factor, viz. a *higher level of the development of the welfare state* (e.g., a highly developed social security system in one country may attract migrants from a country with a less developed system). In the context of European migration this factor may play a role in migration from East to West and to some degree also from South to North.

The choice of migrants in favour of a specific country may next also be the result of *economic dependency or dominance* of the country of origin on the country of destination.

Another factor in the European context may be the existence of *economic and technical assistance programmes* in the country of destination, as it is well known that Western European states offer economic, financial, and technical assistance to members of Eastern European countries as well as to those from developing countries. Such migrants are - even though their migration streams do not play an important role in a quantitative sense - very often the bridgeheads for further migration flows.

*Political dependency and dominance* - as a result of former colonial linkages between some West European States and South-American, African and Asian countries - may also be seen as another critical success absorption factor. Especially in the case of refugees it is noteworthy that they were often attracted by stable democratic systems with more guarantees for social rights, so that essentially the *difference in the internal political system* between the country of destination and origin plays an important role. Given the politically unstable situations in the majority of the East European countries, it seems plausible to state that the political situation in West-European countries will - at least in the near future - be an important absorption factor. Finally, as mentioned already in the discussion of the regulatory approach, a very important factor in the first group of critical success absorption factors is *immigration policies*. Because of the growing importance of the role of policies concerning international migration in Europe, we will discuss this factor separately later on.

### 7.4.2 Mass Culture Connections

The second group of critical success absorption factors refers in particular to *cultural similarities* and *value systems* of the country of origin and destination. The above described European migration patterns - especially the migration between hinterlands and homelands (see Section 7.2) - are at least to some extent the result of this fact.

Similar to this first factor is *cultural dependency or dominance* - the so called 'Westernization', especially in non-European states - as a component of the second group of absorption factors.

The *media diffusion* of cultural life style in television, radio, and press in the country of destination mainly plays a role as an art information source available in the country of origin. With the ever increasing importance of the mass media (especially in the developing world), the importance of information transferred by these media as an absorption factor, is also increasing.

The existence of a *common language, common religion, compatible religious beliefs* as well as *similarities of the educational systems* are further factors attracting and absorbing migrants. However, the latter group of factors clearly illustrates that some factors only play a complementary role next to other factors in the absorption process.

As a last absorption factor *face-to-face contacts through international travel* should be mentioned, especially in light of the growing international mobility.

### 7.4.3 Family and Social Networks

The third group of absorption factors refers to the fact that connections with relatives, friends and former community members in destination countries are also strong elements in prompting a move and facilitating settlement by providing help with jobs and housing. A first factor to be mentioned in this context is the geographic dispersion of relatives and friends, which quite often is characterized as a *geographic clustering of relatives and friends in the country of destination.*

Besides these connections of relatives and friends it is the *connection with members of the home country or community* - who are not necessarily relatives or friends - which plays a major role as social networks and consequently as a critical success absorption factor.

As a third factor we point to the historically grown depth and intensity of family and community relationships. In this context, we stress the fact that the *duration of migration streams* established in the past influences the effects of family and other social connections with the country of destination.

Also *visiting and communication patterns of family and friends* in the country of destination can stimulate new migrants in the country of origin to enter a specific country of destination.

The *social and economic status of previous migrants* and the *frequency and amount of remittances* may give potential migrants in the country of origin important signs and information on the possibilities for foreign migrants in a specific country of destination.

*Normative household and family structures and normative family obligations and commitments* may in a way also be considered as critical success absorption factors. The fact that some cultures place more emphasis on the welfare of members of immediate family (direct kinship), while other cultures stress obligations to virtually all relatives is also important in this context. An interesting difference may then arise, if immigrants accumulate resources in order to expand the inflow of eligible members (leading to an absorption of new migrants) or where resources are accumulated only for the immediate family and therefore do not - at least directly - attract new migrants.

As a conclusion on this last group of absorption factors 'family and social networks', we emphasize here the strategic importance of informal information transfer from the former migrants in the country of destination given to potential migrants in the country of origin, which leads the latter to prefer a specific country to another country. The importance of these kinds of 'personal' information needs to be analyzed in more detail (see for example Brown et al 1981 for internal migration into Columbus, Ohio).

## 7.5. Policy Issues

Finally, we will pay attention to policy questions. Each European state has its own immigration policy. This *diversity* leads to some divergences which are mirrored in the absorption of different types of migrants from different countries. Two classes of such diverse regulations or policies in different European countries affecting absorption may be distinguished, namely those referring to the *stage of entering* a country and on those referring to the *stage of adaption* (see Fawcett and Arnold 1987).

The first class of policies and regulations related to the stage of entering a country contains issues like: regulations and/or policies about *entry conditions and political asylum, family immigration, deportation and categories of immigrants protected from deportation.* For example, regulations on family immigration stress the question, whether immigration is allowed (or admission procedures are necessary) for spouses and children and which conditions have to be fulfilled in addition (see Costa-Lascoux 1990). For a more detailed comparison of the above mentioned regulations and/or various policies between France, Germany, Italy and the Netherlands we refer to Costa-Lascoux (1990) and for a comparison between Germany, Belgium, France, Great Britain and the Netherlands to Groth and Just (1985).

As an example of the second group of regulations concerned with the stage of adaption, the following issues may be mentioned: *political rights for foreigners in comparison to nationals, the possibilities for foreigners to get a working or residence permit, the possibilities to participate in the social security system of a country and general adjustment policies for immigrants.* In order to illustrate the latter - general adjustment-policies, the example of the Netherlands, where positive discrimination and the emancipation of minorities is advocated, may be mentioned in contrast to the example of Germany, which develops vocational education, social aid and the idea of participation in local life (Costa-Lascoux 1990). Integration policies in France in part seek active participation in the national community by encouraging naturalization and not by recognizing ethnic communities (Coleman 1993).

A comparison, for instance, between France and Germany with respect to political rights of foreigners in comparison to nationals shows, that in France only French nationals are entitled to vote and be elected (with the additional regulation that immigrants can be elected by their compatriots as delegates to some municipal councils). In Germany, in principle the same regulation can be observed, with the exception that foreigners have voting rights at communal elections in some municipalities (Costa-Lascoux 1990).

The effect of different nationality rights in Europe may be illustrated by means of the example of Great Britain. In 1983, three different categories of citizenship with British nationality have been created: the British citizenship (which applies to persons born in the United Kingdom itself), the British citizenship of the dependent territories (which applies to persons born in one of the dependent territories) and the British overseas citizenship (which apply to nationals of former British colonies). Only British citizens can reside in the UK and have the right to free circulation and residence in the UK. Also the examples of the ethnic Germans shows that absorption is dependent on the rights resulting from different definitions of a 'foreign' immigrant or a 'national' immigrant (Costa-Lascoux 1990).

Finally, some remarks on convergence emerging in European states have to be added (see also Velling 1993; Van de Kaa 1993; Meijers 1991). Mainly because of increasing flows of immigrants into Europe and the related problems (such as racial trends in the countries of destination, integration problems etc.), in all Western European countries entry conditions for non-Europeans are becoming more restrictive. For EC-Europeans on the other hand migration between the member countries of the EC becomes much easier with the unification of Europe.

Until now migration between EC countries is jointly regulated by the order 90/364/EWG, which restricts the right to stay in a given EC-country only for people who cannot afford their own living costs. In addition, people who stay more than three months in a country need an allowance. With the agreement for the European Economic Area however, the freedom of movement between the EC countries is extended to the EFTA countries. In addition, migration between Scandinavian countries is to be characterized by freedom of movement. Concerning migration from outside the EC, general trends towards a more strictly coordinated immigration policy may be observed, even though it is not yet entirely clear on which level (for example, the European or national level) this coordination should be effectuated. Besides, some bi-lateral and multi-lateral agreements do exist between countries concerning a coordinated immigration policy. One of these agreements is the Schengen Agreement (1990) signed by Germany, France, Belgium, the Netherlands and Luxembourg. The main purpose of this agreement is a gradual abolition of migration controls at the common (internal) borders and free circulation of citizens of countries which are a party in the agreement, although the competence of migration authorities at the supra-national level will increase. A comprehensive Schengen information system will be developed in order to be able to control the external borders effectively. Introduction of a uniform visa and specification of criteria under which an alien can be given permission to enter the joint territory is also laid down in the Schengen Agreement.

In conclusion, regulatory systems will play an increasingly important role in international migration in Europe. Such systems will have a dual function: to encourage economic efficiency through free movement of people in the EC (and associated) countries and to discourage aliens (non-EC members) to enter the EC territory by introducing more strict regulations and absorption mechanism. Consequently, future international migration flows in Europe will be determined by two main factors: socio-economic disparities, both inside the EC countries and between the EC countries and the rest of world, and migration regulations. Thus, socio-political isolation of European citizens will at the end be determined by economic and regulatory conditions.

## References

Bach RL, Schraml LA (1982) Migration, Crisis and Theoretical Conflict. International Migration Review 16:320-341

Bohning WR (1974) The Migration of Workers in the United Kingdom and the European Community. Oxford University Press London

Boisvert CC (1987) Working-class Portuguese Families in a French Provincial Town. Adaptive Strategies. In: Migrants in Europe. The Role of Family, Labor, and Politics. Buechler HCh, Buechler J-M (Eds) Greenwood Press Westport Connecticut, London, New York 61-76

Borjas GJ (1989) Economic Theory and International Migration. International Migration 23:457-458

Borrie WD (1990) The Demographic Consequences of International Migration. In: NIAS Reports. Based on the Proceedings of an International Symposium held at the Netherlands Institute for Advance Study. Wassenaar, September 1990 1-57

Brown LA, Philliber SG, Malecki EJ (1981) Information source usage in the migration decision. Socio-Economic Planning Science 15:321-330

Castles S, Miller M (1993) The Age of Migration. Guilford, New York

Coleman DA (1993) The World on the Move? International Migration in 1992. European Population Conference Geneva (Switzerland), 23-26 March 1993

Costa-Lascoux J (1990) Immigration Policies and the Single European Market. Contemporary European Affairs 3:90-113
Engelbrektsson U-B (1982) The Force of Tradition. Turkish Migrants at Home and Abroad. Acta Universitatis Gothoburgensis, Gotenburg
Fassmann H, Munz R (1992) Patterns and Trends of International Migration in Western Europe. Population and Development Review 18:457-480
Fawcett JT Arnold F (1987) Explaining Diversity. Asian and Pacific Immigration Systems. In: Pacific Bridges. The New Immigration from Asia and the Pacific Island, Fawcett JT, Carino BV (Eds) Center of Migration Studies, New York 453-473
Feithen R (1986) Determinants of Labor Migration in an Enlarged European Community. Working Paper 86/230. European University Institute, Florence
Freeman GP (1986) Migration and the Political Economy of the Welfare State. Annals of the American Academy of Political and Social Science 485:51-63
Hoffmann-Nowotny HJ (1981) A Sociological Approach Toward a General Theory of Migration. In: Global Trends in Migration. Theory and Research on International Population Movements, Kritz MM, Keely ChB Tomasi SM (Eds) The Center of Migration Studies of New York, New York 64-83
Garson J (1992) International Migration. Facts, Figures, Policies. OECD Observer 176:18-24
Gosh B (1991) The Immigrant Tide. European Affairs 6:78-82
Groth A, Just WD (1985) Wanderarbeiter in der EG. Band 1. Vergleichende 52 Analysen und Zusammenfassungen. Matthias-Grunewald Verlag, Mainz
Kaa DJ van de (1993) European migration at the end of history. European Review 1:87-108
Kritz MM, Zlotnik H (1992) Global Interactions. Migration Systems, Process, and Policies. In: International Migration Systems. A Global Approach. Kritz MM, Keely ChB, Tomasi SM (Eds) Oxford University Press, New York
Kulu-Glasgow I (1992) Motives and Social Networks of International Migration Within the Context of the Systems Approach. A Literature Review. NIDI Paper (unpublished)
Meijers H et al (1991) Schengen. Internationalisation of Central Chapters of the Law on Aliens, Refugees, Privacy, Security and the Police. W.E.J. Tjeenk Willink - Kluwer Law and Taxation Publishers
Muus Ph, Cruijsen H (1991) International Migration in the European Community. Two Scenarios. Background Paper. Congress Human Resources in Europe at the Dawn of the 21st Century. Luxembourg, November 1991:27-29
Nijkamp P et al (1991) Services for the Elderly in Europe. HIVA, Leuven
Nijkamp P, Spiess K (1993) International Migration in Western Europe. Current Politics and Economics of Europe. 3:238-262
Penninx R (1986) Theories on International Labour Migration. Between Micro and Macro Analysis. Paper presented at the Xlth World Congress of Sociology, New Delhi, August 1986
Velling J (1993) Schengen, Dublin und Maastricht - Etappen auf dem Weg zu einer europäischen Immigrationspolitik, ZEW-paper (unpublished draft)

Wood ChH (1982) Equilibrium and Historical-structural Perspectives on Migration. International Migration Review 16:298-319
Zlotnik H, Hovy B (1990) Trends in European Migration. What the Data Reveal. Paper prepared for the Symposium on the Demographic Consequences of International Migration. Held at NIAS, Wassenaar, September 1990
Zolberg R (1989) The Next Waves. Migration Theory for a Changing World. International Migration Review 23:403-429

# 8 Barriers and Bridges in Technology Transfer: Perspectives for Border Regions

M. van Geenhuizen

## 8.1 Introduction

In the past several years it has increasingly been acknowledged that technological knowhow is a major asset in the competitive power of a nation and its regions. Accordingly, the quality of the knowledge infrastructure has increased in significance in the general locational profile of regions.

At the same time, companies are facing a progressively high uncertainty and risk. New technologies have pervasively changed advanced industrial economies. For example, information and communication technology, biotechnology and new materials had impacts which went clearly beyond the use of the technology self, such as in new modes of organization of production and new work practices. A crucial development was also the increasing global competition. In addition, the lifespan of products was shortening progressively, and also the complexity of R & D increased in a number of sectors (Van Geenhuizen 1993; Howells 1992). In order to remain competitive, therefore, companies needed and continue to need to increase efforts in raising their technological level. It is recognized that they increasingly satisfy this need by means of external technology sources (Lawton Smith 1993; Tordoir 1993).

Although uncertainty and risk have increased in general, companies may face different degrees of uncertainty dependent upon the newness of their economic activities (sector), as well as their age (Figure 8.1). Regarding the sector, it is commonly acknowledged in product life cycle theory that uncertainty is relatively large in the stage of introduction of new products. This holds true for the technology which is still unstable, and for the market which still has to be created (Rothwell and Zegveld 1985). At the same time, the amount of uncertainty changes with the age of companies. Particularly in the early years, there is a lack of organizational experience (cf. Hannan and Freeman 1989) and a shortage of marketing knowhow.

It is generally agreed that innovative behaviour is unequally dispersed over space, not only because of sectoral or socio-economic variation among companies in different areas but also because of spatial differences in opportunities to reduce uncertainty and spatially discriminating urban and regional policies. As regards the

| | | COMPANY LIFE CYCLE | |
|---|---|---|---|
| | | young | older |
| PRODUCT LIFE CYCLE | introduction | large | intermediate |
| | growth maturity | intermediate | small |

**Fig. 8.1.** A two-way classification of companies in view of uncertainty

latter, it is important to know that in the past decade the scope of regional policy has drastically changed. The Dutch policy for example, shifted from equality to efficiency and an appeal to the so-called self-organizing potentials of regions since the mid 1980s. This new approach emphasized regional endogenous growth. It is increasingly recognized that this type of growth may be strongly supported by the regional scene through benefits from localized collective learning processes (Camagni 1991; Ratti 1992). As a result, information and knowledge are becoming one of the major critical success factors in regional development policies.

One of the components of such regional policies is the identification of potentials of technology transfer between knowledge institutions and the (regional) business world, as well as barriers and bridges to this transfer. Within the framework of regional policy, Transfer Centres and Science Parks are instruments aimed at advancing the specific technology flow between universities and industry. In particular Science Parks are aimed at the 'creation' of a small scale 'incubation milieu', by means of supplying accommodation close to the university.

With regard to regional economic policy, a major new trend should also be mentioned. Various regions with a 'peripheral' location in the national space-economy have recently gained a more 'central' position within a unified Europe. Accordingly, policies for these regions have to be developed within a framework of new potentials for transborder coordination and integration (cf. Ratti and Reichman 1993). However, it should also be emphasized that competition is shifting from the national scene towards the international scene.

This paper sets out the results of a conceptual exploration and results of an empirical analysis. First, the concepts of technology transfer and intermediation will be considered, in particular their structural characteristics and process characteristics (Section 8.2). Barriers to technology transfer will receive particular attention in Section 8.3. Section 8.4 will explore the specific form of technology transfer from universities to the business world. The empirical analysis then, will focus on barriers and bridges experienced in this flow through Transfer Centres (Section 8.5) and Science Parks (Section 8.6) in the Netherlands. In this context one 'peripherally' located university will be considered. This paper will conclude with a summary of the major conceptual and empirical findings, and with some policy implications (Section 8.7).

## 8.2 Characteristics of Technology Transfer

Technology transfer is one part of a chain of interaction processes between technology and economy, following the generation of ideas (creativity) (cf. Andersson 1991). Technology transfer can be defined as the diffusion of the bundle of knowledge which surrounds a level and type of technology (cf. Charles and Howells 1992). This definition is rather broad, including various types of knowledge, various forms of transfer, as well as a variety of organizational and spatial scales.

The types of knowledge have a strong influence on the forms of transfer. In view of this, the distinction is important between embodied knowledge, such as in physical products, plants and equipment, and disembodied knowledge, such as in patents, experience and learning knowhow. It makes sense that information may be articulated so that it is transferable by written documentation and oral teaching. Otherwise (tacit information) it is only transferable by being learned through practice. Based upon the various types of knowledge, five forms of transfer can be identified (Table 8.1). In each of the forms the vehicle of transfer is different, e.g. human beings, written language, spoken language, hardware, and (images of) practical acts.

**Table 8.1.** Forms of technology transfer

| Form (vehicle) | Examples |
|---|---|
| - Human capital transfer | Ideas, skills and routines in employees |
| - Written transfer | Data, manuals, patents, specifications |
| - Oral transfer | Understanding, abilities, competence |
| - Hardware transfer | Devices, equipment, materials |
| - Tacit transfer | Learning by doing and via images |

Basically, two actors are involved in the process of technology transfer, namely (1) the source of knowledge and (2) the receiver of knowledge (Figure 8.2). Intermediary organizations may be active in this process but their involvement is not strictly necessary. The relationship between source and receiver, and the one of the intermediary organization with source and receiver may be very different. The following (interconnected) dimensions are involved in this relationship:

- Strength
- Duration
- Number of sources/receivers.

Concerning different strengths, relationships are usually weak in situations of temporary and incidental transfer or intermediation, whereas strong relationships develop on the basis of permanent arrangements. Important in this respect is also the number of institutions simultaneously active in the process, which may vary from one single institution to an undefined number of them (multiple-linked). A good example of a single-linked intermediary regarding the source, is the academic Transfer Centre because it is exclusively linked to one university. The so-called Innovation Centres in the Netherlands are good examples of multiple-linked actors because they mediate between a large variety of institutions, both on the source and the receiver side. Regarding the dominant orientation, a distinction can be made between source oriented and receiver oriented intermediaries.

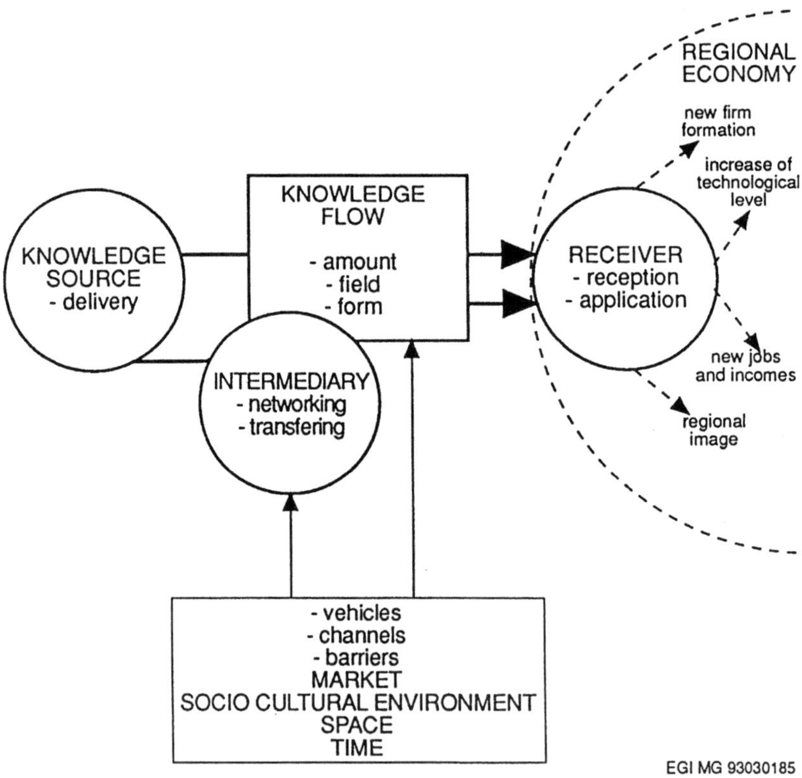


**Fig. 8.2.** Actors in technology transfer

When technology transfer is viewed as a *process*, various stages can be distinguished (Table 8.2). It should be stressed that these stages may overlap to a certain degree whilst various loops may occur. Technology transfer is hardly possible without the first two stages. The actors in the process should know each other and have an appropriate image towards each other. This includes a sufficient level of information about potential supply and demand, and about broad conditions under which transfer may occur, such as cost and form. The next necessary step in the process includes actual contacts between (potential) sources and receivers. Intermediary organizations have prominent roles in these two stages, which can be labelled as *active networking*. A further step in technology transfer includes specific preparatory work in order to guarantee a smooth flow of knowledge. A good example of activities at this stage is the establishment of contractual arrangements and the selection of the most appropriate form(s) of flow.

Within the framework of regional policy, the last stage in technology transfer is extremely important. The primary aim includes the actual use and implementation of the technology by the receiver, in order to raise the technological level of regional business. Other major aims in this respect are to advance the creation of new firms, new jobs and incomes. Building an attractive image as a region or town is also a popular goal of regional (local) policy in this respect.

**Table 8.2.** Stages in technology transfer

| Stage | Characteristics |
|---|---|
| 1. | Building of public relations and image (broad networking) |
| 2. | Establishing contact between specific sources and receivers (focused networking) |
| 3. | Finding forms (vehicles) and channels, as well as removing barriers |
| 4. | Delivery of knowledge |
| 5. | Reception of knowledge |
| 6. | Use and application of knowledge |

In general, various structural characteristics of technology transfer can be distinguished, dependent on the perspective adopted. The market, the social (cultural) environment, space, and time are relevant perspectives here. In a *market* perspective, emphasis is placed upon the price and quality of technological knowledge as a product, the nature of supply and demand and market imperfections, such as the lack of information, etc. This perspective also includes the labour market for academic graduates. Regarded as a *social* phenomenon, technology transfer involves a variety of organizational scales. It operates within and between organizations. It operates also within a variety of scopes, i.e. between the same types of organization (such as universities) and between different types of organization (such as universities and private companies). Within this approach to technology transfer, *network analysis* has increasingly received attention over the past decades (cf. Williams and Gibson 1990). This type of analysis brings to the surface the web of relationships between organizations involved in technology transfer. It focuses, for example, on networks in terms of their size, density, diversity, reachability and dynamics.

Concerning the third perspective, *space*, technology transfer operates on a large variety of scales, ranging from intra-urban to transcontinental. From a spatial perspective, technology transfer can take the form of three networks which are basic to all types of flow, i.e. branched, circuit and barrier networks (cf. Suarez-Villa et al 1992). Technology transfer also has a *time* dimension. Different patterns of transfer can be distinguished in this respect, for example, one singular event and a sequence of events, the latter being divided into regular and irregular patterns. The time dimension of technology transfer depends very much on the nature of the transfer activity. For example, courses and training are usually organized on a sequential basis, whereas the transfer of patents may take place as singular events. Also, the type of channel used in the transfer is important. Insufficient links may prevent transfer during certain times when there is a large demand for communication, such as in telephone networks. With this example, however, our attention is already shifting towards barriers. First, then, it is necessary to consider this concept.

## 8.3 Barriers to Technology Transfer

Barriers to technology transfer may be interpreted as all constraints that impede a smooth flow of information on technology in space and time (Nijkamp et al 1990). Barriers have a disruptive influence on knowledge transfer and are, therefore, indicated by discontinuities in the intensity of this transfer.

The definition used in this article includes however, explicitly a notion on needs or desires for transfer on one or both sides of the potential flow, or among politicians. Thus, there is evidence of barriers when a need for increase of information transfer cannot be fulfilled. When there is no transfer simply because any demand or desire is absent or is decreasing strongly, the question of barriers is not relevant.

Within each of the above indicated 'environments' of technology transfer various potential barriers can be identified (Table 8.3). These barriers are associated with the various stages in the process of transfer as discussed in Section 8.2. An important distinction can be made between general barriers which are *inherent* in the nature of the interactions, and intentional barriers which are *imposed* for reasons of protection against access in the first stages of transfer. A common barrier within the former category is *inertia* of networks to expansion (cf. Håkansson 1988).

From a *market* perspective, transfer may be hampered by a lack of supply or demand. Barriers in the first stages may also be caused by low skill levels on the supply and the demand side of the market, for example, with regard to the image of the actors (cf. Brown 1981). In addition, the nature of the knowledge itself may well contribute to friction on transfer. In innovation diffusion theory, it is emphasized that the higher the level of complexity and cost, the more time it takes before the technology has reached all potential adopters (cf. Davies 1979). A different type of barriers occurs further in the process of transfer, namely as a result of discrepancies between demand and supply in communication channels, i.e. when congestion occurs (Nijkamp et al 1990). Congestion is however, strongly time-dependent due to a fluctuation between peak and off-peak hours. A further important category of barriers from a market perspective aims at protection for economic reasons. For example, patent protection simply forbids others than the patent holder to produce the involved product for a certain number of years. Similarly, tests and certification of new products by importing countries prevent an immediate introduction of new product technology to the market.

Many barriers to technology transfer originate in the *social (cultural) environment*. Language is by far the most important barrier here, preventing networking as well as reception and use of knowledge. Language barriers include spoken, written and computer language, as well as the vocabulary used in communication. The latter barrier type follows, for example, from different stages in the development of the technology (basic and applied) and from differences in the organizational culture between source and receiver (Williams and Gibson 1990). A further category of barriers is concerned with social class differences due to educational and income divergencies or religion. Purposive protection against access to technology is realized by means of keeping group membership exclusive, for example, schools, clubs and societies. These types of barrier already prevent networking in the early stages of the process. With regard to remote regions or islands, social (cultural) barriers have often come into existence because of centuries of isolation from the core-area. This may be reflected in a different dialect and lower educational levels but, more importantly, also in divergent attitudes following from social class differences or ideological differences. Over the past several years, the importance of socio-cultural obstacles has been increasingly emphasized in research on technology transfer (cf. Lundvall 1988; Malecki 1991; Gertler 1993).

A large number of researchers has chronicled the influence of geographical *space* (Charles and Howells 1992), both on the micro-scale (within a building, on a site) and the meso- and macro-scales (regional, national and international context). On the two latter scales, the diffusion of knowledge is thought to follow two spatial patterns, namely contagious and hierarchical diffusion (cf. Hägerstrand 1965; Florax 1992). In the contagious pattern, the diffusion of knowledge is concentrated in the vicinity of the originating source and decays strongly with distance. In the hierarchical diffusion pattern, knowledge transfer is first found in central places whereas it trickles down to places of lower order later in time. From the preceding it follows that large distances and low population densities cause a friction to technology transfer. In addition, various natural barriers should be mentioned. Good examples are mountain areas, islands and areas with unfavourable climatic conditions. Natural access obstacles have however, declined in importance in the previous decades due to tremendous advances in transportation and telecommunication. Social or cultural barriers caused by (former) natural obstacles may, nevertheless, survive for a long time. A further important spatial barrier is caused by political borders. However, in Western Europe we observe that regions situated in border-areas of national economies are now increasingly centrally located within a unified European space.

**Table 8.3.** Potential barriers to technology transfer

| Origin | Type of Barrier |
|---|---|
| **Market** | |
| a. General | Low supply/demand of knowledge |
| | Small skill to identify supply/demand |
| | Small skill to marketing of knowledge |
| | Small skill to use transfer channels |
| | High complexity of knowledge |
| | High cost of knowledge |
| | Unknown knowledge |
| | Congestion |
| | Network inertia |
| b. Protection | Market regulation |
| | Trade borders and (fiscal) tariffs |
| | Secrecy and intellectual property protection |
| | Standardization and certification of products |
| **Social (cultural) environment** | |
| a. General | Language and vocabulary disparities |
| | Educational and income disparities |
| | Cultural behaviour disparities |
| | Network inertia |
| b. Protection | National borders |
| | Social group protection |
| | Political and ideological protection |
| **Space** | Long distance (micro- to macroscale) |
| | Low population density |
| | Natural barriers |
| | Political barriers |
| **Time** | Peak and off-peak hours |
| | Divergent time-zones |

Elsewhere, for example in the former Soviet Union, we observe the emergence of new political borders and barriers. *Time* causes potential constraints in two ways. Within daytime, the transfer infrastructure may not be able to handle all demand during a number of peak hours. This obstacle is most obvious in telecommunication. The second source of time constraints rests on the system of global time-zones. Knowledge transfer may be hampered by a lack of overlap of working times between the two sides of the potential flow. This holds largely for flows by telephone and is clearly visible in world-embracing services such as stock transaction, although lack of overlap is reduced as much as possible (Warf 1989). The above discussion indicates that time barriers usually occur when the first stages of the transfer process have been passed.

To conclude, a large variety of potential barriers hamper technology transfer. Natural barriers, or broadly, spatial barriers seem to have lost importance, whereas evidence on the relevance of socio-cultural barriers increases. The next section will explore specifically the characteristics of technology transfer from universities in order to set a framework for barriers and bridges in this transfer, to be discussed further in this article.

## 8.4 University Transfer

University-industry transfer of technology operates either directly between academics and companies, or by means of intermediary institutions. The general forms of technology transfer, discussed in Section 8.2, are clearly applicable to university-industry flows (Charles and Howells 1992; Dalton 1992). Table 8.4 lists the major forms of university-industry knowledge transfer and a few examples of them by focusing on the one-way flow from university to industry.

**Table 8.4.** Forms of university-industry technology flows

1. *Human capital*: academics part-time appointed in industry; academic graduates working in industry; academics founding a new company based on an innovation (possibly) in Science Parks.
2. *Written documents*: academic invention taken up by existing industry (licensing, patents, specifications); publication in scientific journals, textbooks or manuals; reports on contract research results.
3. *Oral transfer*: academic training, courses, presentation at conferences, consultancy.
4. *Hardware*: supply of accommodation and equipment, such as laboratories, clinical trial facilities, devices, computers and entire information systems.
5. *Tacit*: (images of) acts in laboratory routines, engineering, design, treatment of medical patients.

Since the early 1980s, many universities have intensified their links with industry, and they have also sought new modes of collaboration and intermediation of transfer. Academic Transfer Centres have the primary task of intermediation for transfer, including written, oral and hardware knowledge. Science Parks are largely concerned with transfer by means of human capital as they foster direct formation of companies by academic staff or graduates (spin-offs). They aim particularly at the creation of a physical and organizational 'incubation milieu' in which academic innovations can lead to commercialization.

In the course of time, knowledge transfer has started to constitute a new profession and a new branch of services (Charles and Howells 1992). The major components of these services are listed in Table 8.5. General services reflect the first stages of the transfer process (networking), whereas there is also an emphasis on contractual arrangements. Specialist services are found in agencies with specific additional functions, such as assistance in the design of business plans when there is involvement in attracting academic spin-off. Furthermore, in the United States, patenting and licensing have become a speciality of various university agencies, named Office of Technology Licensign (OTLs) (Parker and Zilberman 1993).

**Table 8.5.** Major components in academic knowledge transfer services

| Type of services | | |
|---|---|---|
| General | - | Internal (university) and external contacts (networking) |
| | - | Missionary work and marketing (networking) |
| | - | Negotiation on behalf of the university or other institute that is represented (finding forms and channels) |
| | - | Drawing up and administering of legal agreements, patents, licences and other forms of contract (finding forms and channels) |
| Specialist | - | Identifying matches between supply and demand of various client groups by means of databases |
| | - | Pursuing opportunities for obtaining grants from private and public bodies |
| | - | Business planning and advice |
| | - | Patenting and licensing |

In the publicity surrounding university-industry links, Science Parks and spin-offs have received a great deal of attention. Science Parks are among the most popular regional development strategies currently in use, in Western Europe, the United States and increasingly, the Pacific Rim. The vast majority of Science Parks has been established since the early 1980s. However, particularly the ones that attracted much attention are clearly older, such as Stanford (1952) and Research Triangle (1959) in the United States, and Cambridge (1972) in the United Kingdom.

In the literature, the term Science Parks is used both in a broad and a narrow sense. In a broad sense it covers a wide range of high-tech developments, including university 'incubator blocks' as well as the Japanese phenomenon of the comprehensive Technopolis. In the present analysis, the term Science Park will be used in a more narrow sense by following the definition of the United Kingdom Science Park Association (cf. Dalton 1992). Accordingly, a Science Park is a property based initiative which:

- has formal and operational links with a university or another higher educational or research institute;
- is designed to encourage the foundation and growth of knowledge based business and related organizations;
- has a management function which is actively engaged in the transfer of technology and business skills to the residents on site.

Despite the above definition, Science Parks may still be different in size, type of accommodation, and mode of management selected to suit the needs of specific situations. A further difference concerns the regional location. Science Parks have been established both in national core-areas and in peripherally located regions. This divergent location may have caused differences in proximity to (other) first order knowledge centres, and also divergences in size of the regional input and output markets. In addition, there may be a diverse benefit from comprehensive regional or urban policies, including for example, investment grants.

Unlike academic Transfer Centres, Science Parks are usually founded by different parties (Van Geenhuizen 1986; Nijkamp et al 1992). Universities and government agencies have been involved in the establishment of the vast majority of Science Parks. The private sector has thus far not been very active in Science Park development. Private firms, which already play a role nowadays, can be subdivided into two major categories, namely real estate developers and investment banks, largely operating for profit (and the latter also for prestige and public relations) (Currie 1985), and large local firms that invest in Science Parks for reasons of public relations and social responsibility.

The principal actors in Science Park development share only a few goals, e.g. regional development and technological development. At the same time, they give divergent priorities to these goals. Such a multiple objective situation causes a limited chance for consensus about the definition of *success* of Science Parks among the principal actors. This observation leads on to the issue of evaluation of Science Parks or other intermediaries in terms of success or failure.

There are many reasons why success or failure cannot be defined easily (cf. Luger 1992). First of all, success is a normative concept, including a notion about a 'standard' which should be satisfied. In addition, when success is defined in terms of for example, net job creation or net increase of university research budgets, the problem of counterfactuality arises. Outcomes are observed where a Science Park is located and a Transfer Centre is operating, while it cannot be known what the outcomes would have been when both did not exist (zero-case). Special evaluation procedures must be applied in order to solve this methodological problem. A further issue is concerned with the time-dependency of success and failure. Due to the changing nature of the processes over time, success must be measured in different ways over the various development stages of Transfer Centres and Science Parks.

In the light of these methodological issues, the following analysis will merely focus on the performance of intermediary organizations and avoid a strict evaluation in terms of success or failure.

## 8.5 Transfer Centres in the Netherlands

The following sections will explore barriers and bridges to technology transfer experienced in the Netherlands. Specific attention will be given to a university located in the Periphery (according to the traditional national perspective), namely Groningen.

Groningen is a medium-sized town (around 170,000 inhabitants) located in the Northeast of the Netherlands (Figure 8.3) (Appendix 8.1). The region of Groningen has been included in regional development programmes for a long time, due to a one-sided economy and relatively high unemployment. Unemployment in the Province of Groningen varies between 13.0 and 16.5% of the labour force, whereas the national average is 9.1% (1990) (Labour Force Survey 1990).

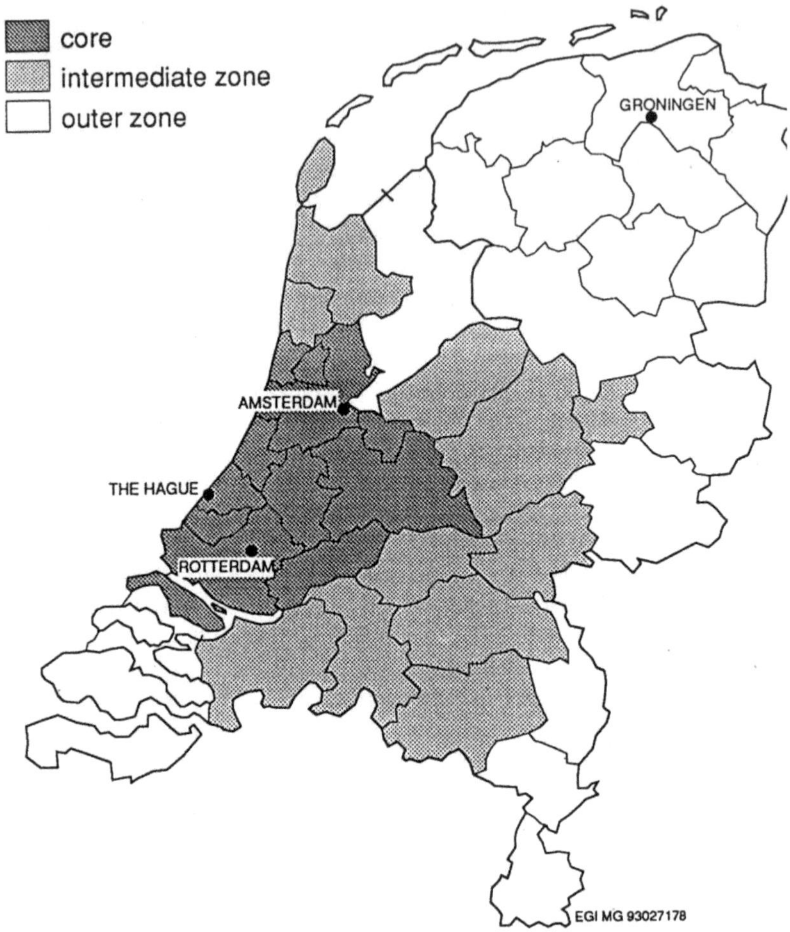


**Fig. 8.3.** The Netherlands in Three Zones

The Dutch system of Transfer Centres was founded in 1981. There are as many Transfer Centres as universities in the Netherlands, namely thirteen (Figure 8.4). In the early years, the system was fully subsidized by the central government. After 1986, however, subsidies were decreased by 25% per year and finally, terminated. This shift in policy caused a rather divergent development of academic transfer organizations in the late 1980s. With regard to (legal) status and type of activities, different models came into being. For example, Groningen remained closely integrated within its university and also became fully financed by this university. On the other hand, a transfer agency with a strong commercial orientation emerged in Tilburg. This agency adopted the company status with limited liability.

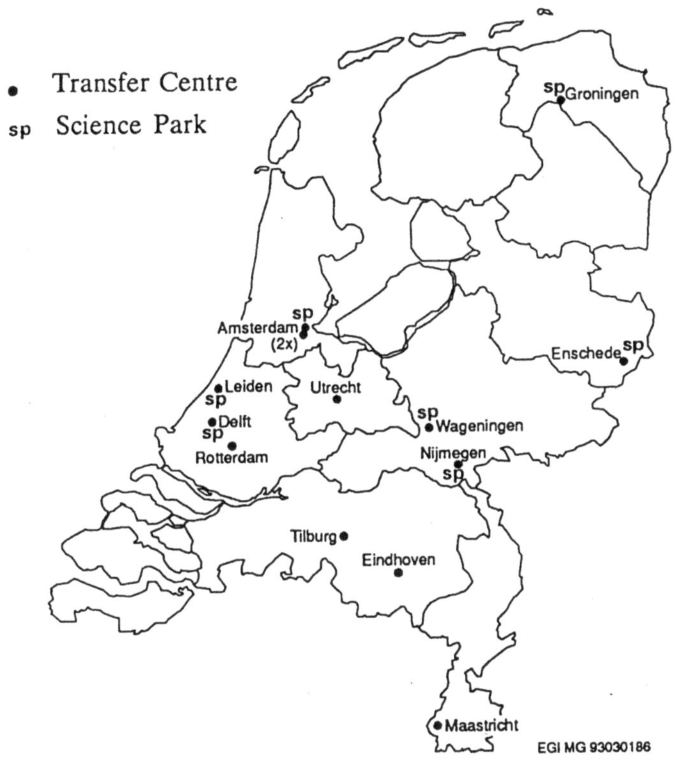


**Fig. 8.4.** University Transfer Centres and Science Parks in the Netherlands

As far as the demand for activities is concerned, Transfer Centres have put a different emphasis on requests by the university for transfer and requests by external target groups. On the whole there was however, a move towards activities initiated by the university, for example, the patenting of university innovations (State University of Leiden 1989, 1990). Thus, university Transfer Centres in the Netherlands have developed a stronger source-orientation in the late 1980s. At the same time, a *downward* trend has become evident in the volume of transfer. This can be illustrated with the number of projects carried out by the Transfer Centre Groningen in the years 1984 to 1991 (Table 8.6). One can observe an initial increase followed by a clear decrease of activity: the number of transfer projects amounted to almost 300 in 1987, whereas it felt down to around 200 in 1991.

As indicated above, this development can easily be explained with the emergence of new types of sources and intermediaries in the market for technical knowledge since the late 1980s. Various higher educational institutes across the country established their own transfer agencies (since 1986) and a system of Innovation Centres also came into being (since 1988). As a consequence, Transfer Centres at a short distance from a new Innovation Centre may have experienced a certain amount of 'catching' of their activities. At the same time, also a number of commercial consultancies started activities in the field of technology transfer. It can be concluded that the emergence of new sources and intermediaries has led to a loss of potential activities among academic Transfer Centres. Further consequences may have been the rise of missing links between the different transfer networks, as well as a decrease of transparency of the supply structure towards potential receivers.

**Table 8.6.** Number of projects carried out by the Transfer Centre Groningen

| | Absolute | Index Number |
|---|---|---|
| 1984 | 231 | 100 |
| 1985 | 269 | 116 |
| 1986 | 283 | 123 |
| 1987 | 297 | 129 |
| 1988 | 239 | 103 |
| 1989 | 239 | 103 |
| 1990 | 245 | 106 |
| 1991 | 193 | 84 |

Source: Annual Reports

In expanding knowledge networks from universities, a large degree of *inertia* is generally encountered, resting on the largely stable and conservative nature of contact networks (Charles and Howells 1992). This phenomenon can be illustrated with the establishment of new contacts by the Transfer Centre Groningen (Table 8.7). The most important basis for new contacts rests in existing relationships. In addition, many new contacts are based upon recommendation given by various other parties. Accordingly, the vast majority of all new contacts (75%) follows from previous knowledge about the Transfer Centre.

**Table 8.7.** Basis of contacts with the Transfer Centre Groningen (1990)

| | Percentage | Ranking |
|---|---|---|
| *Based on previous knowledge* | | |
| Previous customer's relationship | 47 | 1 |
| Recommendation by third party a. | 28 | 2 |
| *Based on new knowledge* | | |
| Presentation on conferences, fairs | 13 | 3 |
| Otherwise (advertisement etc.) | 12 | 4 |
| Totals | 100 | |

Source: Annual Report 1990. a. Personal recommendation included.

**Table 8.8.** Geographical spread of customers of the Transfer Centre Groningen (percentage shares)

| Location | 1984 | 1987 | 1990 | 1991 |
|---|---|---|---|---|
| Internal a | 10.8 | 9.1 | 7.8 | 7.8 |
| Groningen Province | 43.3 | 52.9 | 56.3 | 56.0 |
| Remaining North b | 32.5 | 24.9 | 20.8 | 20.2 |
| Remaining country | 13.4 | 12.1 | 14.3 | 14.5 |
| Abroad | 0.0 | 1.0 | 0.8 | 1.6 |
| Totals | 100 | 100 | 100 | 100 |

a Within the university or Academic Hospital. b Provinces of Friesland and Drente.
Source: Annual Reports

**Table 8.9.** Potential barriers to academic technology transfer

| Intermediary | Type of barrier |
|---|---|
| **Transfer Centre** | Low university potentials for commercialization<br>Competition from new sources/intermediaries<br>Missing links between various sources/intermediaries<br>Lack of transparency of supply structure for potential receivers<br>Inertia of contact networks<br>Lack of knowledge about and appropriate image of university potentials among receivers |
| **Science Park** | *University:*<br>Low potentials for commercialization<br>Lack of contractual arrangements which advance a shift of academic staff to business<br>Low responsiveness and flexibility in supply of supportive services<br><br>*Science Park organization:*<br>Inappropriate accommodation for spin-off<br>Accommodation types which are plentifully offered elsewhere in the region<br>Target firms which are not appropriate for potentials of the region<br>Lack of supportive services<br>Management 'gap' after initial success<br>Relaxation on selection criteria for future occupants<br>Low level of coordination between various actors (founders)<br>Passive role in identifying potentials for commercialization |

Despite the fact that new contacts are mainly based on previous knowledge about the Transfer Centre, there is no evidence of a spatial inertia of the contact network (Table 8.8). Since the mid 1980s, there has been a clear shift from customers in the remaining Northern Provinces to customers in the Province of Groningen. In this respect the network has 'shrunken' considerably. At the same time a small increase of contacts outside the North (foreign countries also included) has become evident.

A further type of barriers to technology flow from universities can be summarized as *mental* barriers. These include for example, a lack of invitation to participate, or a lack of awareness among firms about the potentials for collaboration (Charles and Howells 1992). Regarding the inert character of networks, this means that very strong efforts are needed in missionary work and marketing in order to increase the size of transfer networks.

Table 8.9 summarizes the major barriers to technology transfer from universities, discussed so far. It also summarizes barriers to the development of Science Parks, particularly the growth of academic spin-off which will be discussed in the next section. Within this framework the Science Park in Groningen, Zernike Science Park, will serve as an example of a successful avoiding of various barriers.

## 8.6 Science Parks in the Netherlands

Currently there are seven Science Parks (in the strict sense) in the Netherlands (Table 8.10, Figure 8.4). They appear to be very different in age and size. The oldest and far the largest is in Enschede (106 establishments and 1,115 jobs after twelve years). The Science Park in Leiden is also relatively old (nine years) but slightly less successful in attracting a large number of establishments (twenty-five). However, it has been successful in attracting relatively large companies, particularly from abroad. Considering its age, Groningen performs relatively well (fifty-one firms and 460 jobs after five years).

Apart from the number of companies and their employment on Science Parks, the type of relationships of these companies with the university is also of interest, regarding the overall aim of facilitating knowledge transfer. In the Netherlands, a large differentiation is evident between companies on Science Parks (Bartels and Wolff 1993). From all types of knowledge relationships with universities, the two most important seem to be the recruitment of staff (not specifically from the nearby university) and the use of facilities of the nearby university. An important position is also held by the (temporary) employment of company personnel at the university and the attendance of academic courses, meetings, etc. In sum, although technology transfer with the nearby university can be extremely manifold and intense, this is certainly not the actual overall picture. In this respect, a certain scepticism is needed. This may also hold for the emergence of local networks between firms on the same Science Park. In Sophia Antipolis for example, it was found that innovative local networks were virtually non-existent (Longhi and Quéré 1993). Horizontal modes of cooperation between firms were absent here, due to the fact that the products of these firms were too diverse. Also, there was a lack of trust between innovative large firms and local research institutions. Observations like these raise the question as to whether it is altogether possible to *create* an 'incubation milieu', in the sense of innovative knowledge networks.

**Table 8.10.** Science Parks in the Netherlands (1993)

| Town | Starting year | Size (ha) | Establishments | Jobs |
|---|---|---|---|---|
| Enschede | 1981 | 18.5 | 106 | 1,115 |
| Leiden | 1984 | 30 | 25 | 640 |
| Groningen | 1988 | 60 | 51 | 460 |
| Nijmegen | 1989 | 1.5 | 25 | 100 |
| Wageningen | 1989 | 5.5 | 39 | 400 |
| Amsterdam | 1991 | 20 | 20 | 150 |
| Delft | 1992 | 30 | 15 | 27 |

In view of obstacles to Science Park development regarding the establishment of firms, a distinction can be made between the role of the university and the one of the Science Park organization, the latter being largely involved in set-up and management (cf. Gibb 1985; Van Geenhuizen 1986; Nijkamp et al 1992). The actors involved in Zernike Science Park (Groningen) have clearly been aware of a number of obstacles and organized the Science Park accordingly (Appendix 8.2). As a 'second generation park', it could benefit from the experiences of older ones. In the early years, emphasis was placed on the fostering of commercial projects and start-up firms and later, priority was given to real-estate development (De Lange 1992). The vast majority of its current occupants (almost 90%) is a spin-off from the local university or related institutes.

In terms of the role of the university, it is obvious, but sometimes overlooked, that there should be in-house expertise which has strong potential for commercialization. Universities with a weak orientation on for example, applied informatics, biotechnology and advanced materials are less favourable for technology transfer to industry than other universities. In addition, a lack of inviting arrangements may hamper a smooth shift of potential academic entrepreneurs to (part-time) involvement in commercial ventures. A major component in this respect is the creation of attractive legal and contractual conditions for a partial shift of university staff towards their own business. In Groningen, a number of contractual agreements are being developed between university departments and start-up companies, concerning salary cost and share ownership (De Lange 1992). A factor which may in general discourage the rise of spin-off firms, is low flexibility and responsiveness of universities in the dealings with them. This also concerns the use of university facilities such as computers, laboratories and their equipment.

With regard to the Science Park organization, a useful distinction can be made between the creation of two types of 'seedbed' conditions, namely physical factors (accommodation) and organizational factors. Regarding the former, it is of utmost importance that small and cheap accommodation is available for rent, usually in 'incubator blocks'. These units are preferably flexible in order to provide room for potential expansion, which is, for example, particularly important for new biotechnology ventures and their laboratories (Feldman 1986). Also, the type of rent contract should give some flexibility to starting companies. A very important further consideration regarding the type of accommodation, is concerned with the nature of the regional (local) market for business accommodation. From experiences in Britain it appears that an over-supply of accommodation of the same type within one region may hamper the growth of Science Parks (Van Geenhuizen 1987). One related point for attention is the type of target companies (and concomitant accommodation). For example, in remote regions the large distance to an international airport may prevent certain internationally oriented companies from establishing. It is advisable, therefore, to exclude accommodation for such companies on Science Parks in remote regions.

Regarding organizational factors, first of all, a range of supportive services needs to be made available to start-ups, such as advice in firm foundation, management and marketing advice, as well as financial support services (venture capital). Firms on Science Parks typically face a large amount of uncertainty, due to their new technologies and young age (Section 8.1). Therefore, a strong focus of the Science Park organization on advice and problem-solving is very important, although it should be stressed that problems faced by Science Park occupants are not the same for all categories. For example, foreign subsidiaries encounter small barriers while independent start-ups usually face a large number of difficulties. In particular, independent start-ups in modern biotechnology are very vulnerable, due to a high investment level (advanced laboratory equipment regarding the practice of DNA-technology and the use of radio-nuclides). At the same time, specific problems are faced in the often global marketing (Daly 1985; Hagedoorn and Schakenraad 1990).

Zernike Science Park in Groningen has two major tools available to foster new high-tech enterprises (De Lange 1992). It actively coordinates and manages projects which are expected to result in commercial products, processes, or new companies. In view of this, it also supplies training for management and access to venture capital.

A further point of concern is the quality of management in the long-term. The success of Science Parks in the first years usually rests on the shoulders of one or two strong managers. It is important to realise that these managers are often 'captured' by other organizations after the success of the park has become evident. Therefore, it is advisable to pay close attention to the succession of the first manager, in order to prevent the danger of a management 'gap' after some years.

A factor which is often overlooked in Science Park development, concerns the consistency in the selection of occupants of the park. The history of older parks indicates that the concept of 'seedbed' for high-tech business is in many cases relaxed so that general business or low-tech manufacturing is coming in, in order to maintain the park's viability as a real estate project. With regard to the United States, Luger and Goldstein (1989) found that only around 25% of the parks continued to exist according to original plans. The risk of a future shift in type of occupants, however, may decrease the attractiveness of the Science Parks in terms of the preservation of their image and the continuation of their (specific) service supply. This may particularly be true for so-called theme-parks, such as those regarding bio-medical technology and agro-technology. A further barrier to growth of Science Parks is strongly connected with the different actors involved in the establishment (Section 8.4). From the beginning, constructive ties should be kept between them in order to prevent the development of too divergent policies.

With regard to spatial barriers, the performance of Zernike Science Park in Groningen indicates that a remote or peripheral location is certainly not a major obstacle to Science Park development, and when such obstacles do exist, they may be 'compensated' for by means of a strong organization. The specific organizational strength of Zernike Science Park includes a comprehensive approach to firm formation, strong ties with other responsible parties, a strict application of the original concept, an active role in the identification of academic potentials for commercialization, and a strong commitment of the organization's managers.

## 8.7 Conclusion

The availability of technology in a region is increasingly regarded as a crucial asset to economic development. Within this framework, this has paper explored the concepts of technology transfer and intermediation, their features as a process and their structural characteristics. By conceiving technology transfer as a process, six different stages could be distinguished, starting with public relations (including image building) and ending with the application of the technology by the receiver. Intermediary organizations have prominent tasks in the early stages of technology transfer, i.e. in active networking.

Barriers have been defined in this paper as all constraints that prevent a smooth flow of information. Various structural barriers to technology transfer could be identified theoretically by adopting four different perspectives, i.e. the market, social (cultural) environment, space and time. Spatial barriers, particularly a small natural access, have increasingly lost importance, whereas the related social (cultural) barriers still exert a significant influence on technology flows. Language constitutes the most significant barrier at work in all stages of the transfer process. This also includes the vocabulary used in the organizational cultures of source and receiver. Unlike language, purposive protection against access to technology has its major impact in the first stages of transfer.

This paper also focused on empirical findings in the Netherlands about various intermediaries. The results of this part led to an inventory of barriers to transfer from universities (Table 7.9). This inventory can serve as a 'checklist' in future decisions on academic Transfer Centres and Science Parks. Regarding Transfer Centres, various barriers have recently arisen due to the emergence of new knowledge sources and new intermediaries. These barriers include a loss of receivers, (potentially) missing links between networks, and small transparency of networks. In addition, structural barriers involve a lack of awareness and appropriate images (mental barriers), and resistance to new contacts (inertia), all at work in the early stages of the transfer process. In light of this, it is advisable to target policy efforts on the following:

- To remove mental barriers by advancing intense networking between universities and small and medium-sized firms, particularly by increasing the accessibility of universities to these firms.
- To investigate the demand for technology transfer through various intermediaries and, when necessary, to investigate the best way to 'bundle up' a number of these intermediaries.

This paper has also examined university Science Parks in detail. The barriers in this respect are structurally different from the ones discussed above because the transfer involves largely human capital, in the form of firm foundation and employment by (ex)academics. The growth of academic spin-off may be hampered by low commercialization potentials of the university and a lack of attractive conditions for academics to start their own business. A major critical success factor is the supply of 'customized' accommodation and supportive services to young firms. It remains however, questionable whether the creation of an 'incubation milieu' in the sense of a localized knowledge network is a realistic aim.

This study has taken Groningen, located in the Periphery of the Dutch space-economy, as an example. It appeared that technology transfer from universities in remote regions to the business world is not structurally different from technology transfer in Core-regions. The Zernike Science Park illustrated that Science Park development is quite possible in more remote regions. Spatial barriers, if any, can be compensated for by the quality of the Science Park organization and management. In addition, it should be emphasized that many 'peripheral' locations in the *national* economic systems are currently transforming into 'central' locations within the *European* space-economy. Accordingly, universities in border areas are facing new opportunities for transborder networking and for knowledge transfer to business related to new functions of their region, such as transit and 'gateway' functions.

## References

Andersson ÅE (1991) Creation, innovation, and diffusion of knowledge. General and specific economic impacts. In: Sistemi urbani 1-2-3:5-28

Bartels CPA, Wolff JWA (1993) Science parken in Nederland. In: Economisch Statistische Berichten 10 November 1993:1038-1041

Brown L (1981) Innovation Diffusion, A new perspective. Methuen, London
Camagni R (1991) Local 'milieu', uncertainty and innovation networks. Towards a new dynamic theory of economic space. In: Camagni R (Ed) Innovation networks. Spatial perspectives. Belhaven Press, London 121-144
Charles D, Howells J (1992) Technology transfer in Europe, public and private networks. Belhaven Press, London
Currie J (1985) Science Parks in Britain.Their Role for the Late 1980s. CSP Economic Publications, Cardiff
Dalton IG (1992) Science Parks, a mechanism for technology transfer. In: Topos Review of urban and regional studies May 92:55-70
Daly P (1985) The Biotechnology Business. A strategic analysis. Frances Pinter, London
Davies S (1979) The diffusion of process innovation. Cambridge University Press, Cambridge
Dunford M (1992) Technopoles: research, innovation and skills in comparative perspective. In: Topos Review of urban and regional studies May 92:29-54
Feldman MMA (1985) Biotechnology and local economic growth: The American Pattern. In: Hall P, Markusen A (Eds) Silicon Landscapes. Allen & Unwin, Boston 65-79
Florax R (1992) The University. A Regional Booster? Avebury, Aldershot
Geenhuizen MS van (1986) Science Parks, Prospects for Success? Papers and Proceedings of International Congress Malmö Sweden. International Federation for Housing and Planning, The Hague
Geenhuizen MS van (1987) Science Parks, een eerste inventarisatie. In: Stedebouw en Volkshuisvesting 1987:5
Geenhuizen MS van (1993) A longitudinal analysis of the growth of firms. The case of the Netherlands. Erasmus University (PhD Thesis), Rotterdam
Gertler MS (1993) Being There. Proximity, organization, and culture in the development and adoption of advanced manufacturing technologies. Paper presented at the Annual Meetings of the Association of American Geographers. April 1993, Atlanta
Gibb JM (Ed) (1985) Science Parks and Innovation Centres. Their Economic and Social Impact. Elsevier Science Publications, Amsterdam
Hagedoorn J, Schakenraad J (1990) Interfirm partnerships and cooperative strategies in core technologies. In: Freeman C Soete L (Eds) New explorations in the economics of technological change. Pinter, London 3-37
Hägerstrand T (1965) Aspects of the Spatial Structure of Social Communication and the Diffusion of Information. In: Papers of the Regional Science Association, 16:27-42
Hakansson H (1988) Industrial Technological Development, A Network Approach. Croom Helm, London
Hannan MT, Freeman J (1989) Organizational Ecology. Harvard University Press, Cambridge, Mass
Howells J (1992) Pharmaceuticals and Europe 1992: the dynamics of industrial change. In: Environment and Planning A 24:33-48
Labour Force Survey (1990) Netherlands Central Bureau of Statistics, Voorburg

Lange L de (1992) Establishing a Science Park in the Netherlands. The case of the Zernike Science Park. Paper presented at the Conference of the International Association of Science Parks in Groningen, 5-7 February 1992

Lawton Smith H (1993) Externalization of Research and Development in Europe. In: European Planning Studies 1:465-482

Longhi C, Quéré M (1993) Innovative networks and the technopolis phenomenon. The case of Sophia Antipolis. In: Environment and Planning C 11:317-330

Luger M, Goldstein HA (1989) Research (Science) Parks as Public Investment. A Critical Assessment. Wirtschaftsuniversität, Vienna

Luger M (1992) Methodological Issues in the Evaluation of US Technology Parks. In: Bozzo U, Elias de Freitas J, Higgins T (Eds) Proceedings of the International Workshop on Science Park Evaluation. Commission of the European Communities, Luxembourg

Lundvall BA (1988) Innovation as an interactive process. From user-producer interaction to the national system of innovation. In: Dosi G (Ed) Technological Change and Economic Theory. Frances Pinter, London 349-369

Malecki EJ (1991) Culture's Influence on Technological Development and Economic Growth. Paper prepared for the Conference of the Commission on Industrial Change of the International Geographical Union. Penang, Malaysia, August 1991

Nijkamp P, Rietveld P, Salomon I (1990) Barriers in spatial interactions and communications. A conceptual exploration. In: The Annals of Regional Science (1990) 2:237-252

Nijkamp P, Oirschot G van, Oosterman A (1992) Technopolis and spatial development. In: Topos Review of urban and regional studies, May 92, 93-106

Parker D, Zilberman D (1993) University Technology Transfers. In: Contemporary Policy Issues XI:87-99

Premus, R. (1986) The strategic role of university Science Parks in the high-tech strategies of state and local governments in the United States. Seminar on Science Parks and Technology Complexes in relation to regional development. Venice, June 1986

Ratti R (1992) Innovation Technologique et Developpement Regional. Instituto di Recerche Economiche, Bellinzona

Ratti R, Reichman S (Eds) Theory and Practice of Transborder Cooperation. Helbing & Lichtenhahn, Basel/Frankfurt am Main

Rothwell R, Zegveld W (1985) Reindustrialization and Technology. Longman, Harlow, Essex

State University Leiden, Annual Report 1989, 1990. Leiden

Suarez-Villa L, Giaoutzi M, Stratigea A (1992) Territorial and border barriers in information and communication networks. A conceptual exploration. In: Tijdschrift voor Econ. en Soc. Geografie 83:93-119

Tordoir PP (1993) The Professional Knowledge Economy. University of Amsterdam (PhD Thesis)

Transfer Centre State University Groningen, Annual Report 1984-1991. Groningen

Warf B (1989) Telecommunications and the globalization of financial services. In: Professional Geographer 41:257-271

Williams F, Gibson DV (1990) Technology Transfer. A Communication Perspective. Sage, Newbury Park

## Appendix 8.1

## Location of Groningen in the Netherlands

For Dutch standards, the distance between the Core-area where the major cities are located, and Groningen is rather large. There are no flights from Amsterdam International Airport to Groningen (Eelde), but there are regular flights from Rotterdam Airport. These have a frequency of three times a day and a duration of 45 minutes. By train, Groningen is directly connected with the major cities in the Core-area. This intercity train runs once an hour from Amsterdam, and once in two hours from Rotterdam and The Hague. Travelling time is around two and a half hour. As concerns road traffic, Groningen is at a distance of around 220 kilometres from Amsterdam and 260 kilometres from Rotterdam and The Hague.

**Traffic infrastructure between Groningen and the Dutch Core-area**

| | Connection | Frequency/ distance | Travelling time/ |
|---|---|---|---|
| Air | Rotterdam | 3x per day | 0.45 minutes |
| Intercity train | Amsterdam | 1x per hour | 2.20 minutes |
| | Rotterdam | 1x per two hours | 2.35 minutes |
| | The Hague | 1x per two hours | 2.40 minutes |
| Highway | | | |
| | Amsterdam | 220 kilometres | |
| | Rotterdam | 260 kilometres | |
| | The Hague | 260 kilometres | |

## Appendix 8.2

**Key features of Zernike Science Park in Groningen**

| Feature | Comment |
|---|---|
| Year of foundation | 1988a |
| Concept | - Supportive organization for academic spin-off<br>- In a later stage, priority given to real-estate development. |
| Organizational features | - Comprehensive approach (accommodation, venture capital, business training etc.)<br>- Own tools to advance the start-up of firms<br>- Strong ties between responsible institutions (Zernike Foundation, University, Regional Development Agency, Northern Provinces etc.)<br>- Large commitment of key managers.<br>- Active screening of university on potentials for commercialization |
| Accommodation | 60 hectare ground<br>10.000 m2 incubator floor |
| Number of firms | 51 |
| Type of firms regarding origin | Relocated from elsewhere: 7%<br>Subsidiary (foreign): 4%<br>Academic spin-off: 89% |

a Start as an organization in 1984.
Source: Interview

# Part B

# Empirical Studies on Information and Transport Networks in Europe

# 9 Access to Telecommunication Networks: Regional Variations in Consumption Network Externalities

R. Capello and P. Nijkamp

## 9.1 Introduction

In recent years much attention has been devoted to the diffusion and adoption processes of new information and communication technologies (ICTs). The interest in these technologies is originating from the importance they have assumed in the process of defining competitive advantages among firms and comparative advantage among regions. It is in fact a common idea among economists and policy makers that the diffusion of these technologies are of strategic importance for the economic development of less favoured regions. For this reason in 1987 the EC launched a five year programme, called the STAR Programme, for the implementation and diffusion of these technologies in Objective 1 regions of the Community.

The degree of success of these interventions is related to the diffusion process that these technologies follow in the first phases of their implementation. Under this perspective, it is extremely interesting to define which are the mechanisms better sustaining the diffusion of these technologies among users. The aim of the present paper is to offer empirical evidence on the diffusion (and adoption) processes of information and communication technologies. A new concept has recently been introduced by industrial economists as the main explanation of diffusion processes of interrelated technologies, i.e. the concept of network externality[1]. The basic idea at the basis of all studies dealing with network externalities is that the rate of growth in the demand for interrelated technologies is dependent on the number of subscribers or clients already using that specific technology. This mechanism of interdependent preferences among adopters has been labelled a *consumption network externality,* since, as the traditional economic

[1]This is a result of a broader strategic research project, in which the primary subject of study is the phenomenon of *network externality,* and its effects on the performance of firms and regions, undertaken by authors, see, among others, Capello 1994; Capello and Nijkamp 1994.

theory emphasises[2], a consumption externality takes place when the behaviour of one adopter of the network affects the consumption of another adopter. This concept has to be kept separate from the so-called *production network externality* concept: in this case, the behaviour of one adopter in joining a network affects the production of another adopter: the effects of the adoption can be measured on the production function of other adopters and for this reason it is labelled 'production network externality'[3].

From this simple, yet fundamental, observation a series of theoretical and empirical analyses have followed, trying to conceptualize and prove empirically the significance of network externalities. Up to now, the concept has been widely studied and applied, especially with a view to measuring the impact that it has on the utility function of each subscriber or client[4]. In essence, the basic concept stems from the strong interdependence which exists in the utility function of each potential subscriber to join the network, making the decision highly dependent on the behaviour of others.

The present paper provides a behavioural analysis of the importance of network externalities on diffusion processes in the case of two different economic environments in which these technologies are developed. The STAR Programme provided an interesting area of analysis for less developed regions. However, for our purposes of measuring network externality effects on the willingness to adopt in different economic environments, an analysis developed in a rather contrasting economic milieu was necessary. As a developed region we chose the *North of Italy*, representing a contrasting case with respect to the South of Italy. In fact, the diffusion process of new technologies is more advanced; firms are stimulated by a more developed economic environment, and, last but not least, these technologies, although public technologies, are commercial technologies. The contrasting characteristics undoubtedly, give great emphasis to a comparative study, since they allow the measurement of network externalities as mechanisms for technological adoption in different stages of the diffusion process. In the case of Northern Italy, the development of these technologies is much more advanced than in Southern Italy. Thus, the analysis in both geographical areas will allow to test the importance of network externalities as reasons for adoption in different stages of diffusion.

Some interesting results stem from the analysis. Especially, what is interesting is the high level of similar replies obtained within the two macro-areas, mirroring a common behaviour by firms located in the same economic environment, as we will see in the next section.

---

[2]For a review of the traditional literature on externalities, see, among others, Mishany 1971; Cowen 1988.

[3]For a broad analysis of consumption and production network externalities, both at the theoretical and empirical level, see Capello, 1994; Capello and Nijkamp, 1994.

[4]The literature on network externalities is rather new. See among others Allen 1988 and 1989; Antonelli 1989, 1990 and 1992; Bental and Spiegel, 1990; Cabral and Leite, 1989; Curien and Gensollen, 1987; David 1985 and 1992; Falloner and Saloner, 1985 and 1986; Hayashi, 1992; Katz and Shapiro, 1985 and 1986; Rohlfs, 1974.

In the first part of the paper the STAR Programme will be presented (Section 9.2). In the second part a behavioural analysis of the reasons for adoption of these technologies among a sample of firms (potential adopters) will be presented (Section 9.3) and the differences among developed and less favoured regions will be highlighted thanks to the fact that the analysis has been run in both the North and in the South of Italy. The paper ends with some concluding remarks.

## 9.2 Telecommunications and Regional Development: the STAR Programme

The increasing awareness of the importance of telecommunication infrastructures and services for encouraging economic development has been a prominent feature of the 1980s, as is witnessed by an increasing body of economic literature underlining the importance of telecommunications technologies for economic growth at both a theoretical and an empirical level. This increasing awareness has also stimulated policy interventions in this direction, by interpreting infrastructure in general, and telecommunications technologies in particular, as a power source of economic growth.

The Special Telecommunication Action for Regional Development (STAR) is a European Community Programme launched in 1987, reflecting the general positive attitude towards telecommunications infrastructures as the key elements in the pursuit of regional development. Under the EC auspices, the STAR Programme aims to encourage economic development in the less favoured regions of the Community by means of easier and quicker access to advanced telecommunications technologies. The programme was launched in all the less favoured regions of the Community, the so-called 'Objective 1" regions, namely part of Italy, Spain, Great Britain, France, and the whole of the Republic of Ireland, Greece and Portugal.

The backwardness of the less favoured regions in telecommunications infrastructure and services was the second reason for this direct intervention by the Community. In 1987, when the STAR Programme was launched, the distribution and quantity of the telecommunications infrastructure in those regions was rather poor. The telephone network density was extremely low in comparison with core regions, and the situation was even worse for advanced telecommunications networks and services.

The intervention of the Community aimed at diminishing the disparity in terms of network accessibility between the advanced and less-favoured regions, by installing basic infrastructures (optical fibre networks, satellite networks, ISDN, advanced networks) for advanced telecommunications services and by encouraging the supply and demand for such services. Thus, the STAR Programme is a typical supply driven innovation process, aiming at:

- *developing a technological supply* in areas where the market rules would have never encouraged public operators to make sufficient financial efforts. In fact, the lack of an explicit demand for advanced telecommunications services in these regions does not in fact justify the financial effort required to produce these advanced technologies. However, the project of implementing advanced telecommunications networks has been developed with the full involvement of the public operators for different reasons. First of all, the institutional monopolistic position of most of the European public operators (apart from the British case) would otherwise have meant that it was not possible to realize a full implementation of the Programme. Secondly, the 'additionality' principle governing the European Structural Funds would otherwise not have been fulfilled: the Community in fact expects, as a basic principle for its local policies and interventions, the same financial effort from the National bodies in areas where this kind of investment would otherwise never take place;

- *promoting advanced and innovative telecommunications services,* with the help of pilot services in specific micro-areas;
- *supporting and stimulating a demand for these services,* in order to create a real market at the end of the five-year programme. For this reason demonstration and promotion centres have been developed in order to facilitate the linkage between technological potentialities of the supply and the real needs of the demand. In other words, these centres play the role of intermediaries in the market, with the aim of stimulating a sustainable innovation process. While the promotion of advanced services is based on a sample of firms, this phase is directed to all potential business users.

In concrete terms, the STAR Programme has developed:

- a broadband network;
- experimentation of the Integrated Service Digital Network;
- specialized networks (Itapac, digital switching network, digital circuits, fibre optic networks);
- a satellite network to support the development of advanced services in specific demonstration centres;
- other more specific interventions in some areas;
- five specialized centres for assistance and training in the use of new telecommunications services (such as video conferencing; data banks; on-line CAD/CAM technologies, etc.)

A large group of SMEs in the South of Italy was chosen through an ex-ante evaluation of the potentially most innovative firms in the South of Italy, based on indicators of their technological and innovative development pattern and their entrepreneurial capabilities. These firms were given an on-line connection to the advanced telecommunications services available at the demonstration centres, and sometimes access to the advanced networks (when ready). All services and on-line connections were free of charge. Thus, these firms could use CAD/CAM systems on their (on-line) personal computer, could have access to data banks (different for the different industries) and could exploit the electronic mail service.

In order to develop all networks and services mentioned above, the Programme has been organized in two large projects aiming at:

- developing advanced telecommunications networks (fibre optic networks, Integrated Service Digital Network, digital telephone networks) for the

development of advanced services (the so-called 4.1 measures)[5]. Unfortunately, these networks still have a very low level of connection with networks in the North;
- developing promotion and demonstration centres (the so-called 4.2 measures).

The structure of the Programme is presented in Figure 9.1. The Programme was run by DG.XVI (Regional Policy) of the EC with a small peripheral role played by DG.XIII (Science and Technology). 50% of the available funds were spent by DG.XVI for both the 4.1 measures (infrastructure provision) and 4.2 measures (service promotion).

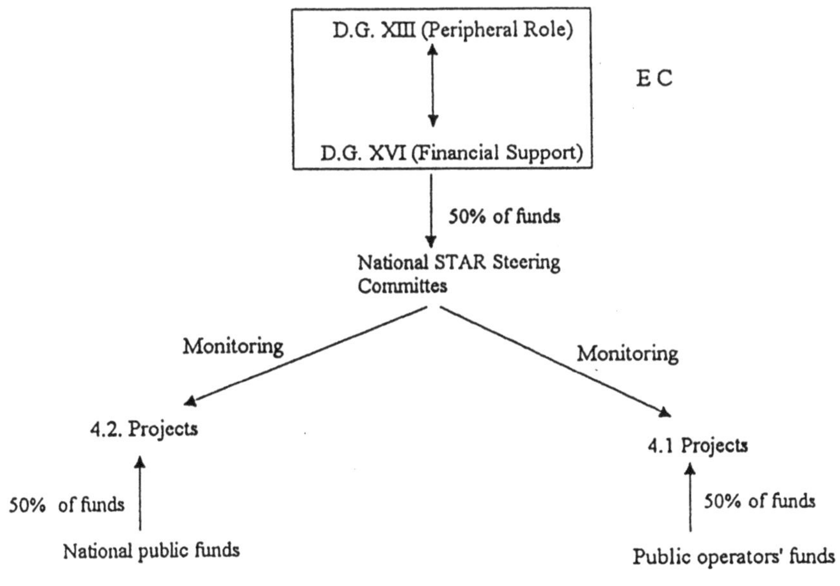


**Fig. 9.1.** Structure of the STAR programme

A National STAR Steering Committee has been created for monitoring both 4.1 and 4.2 actions. In the case of service provision, the National Committee was in charge of the decision on the kinds of projects to be developed under the 4.2 measures, which had to be additionally financed with national public funds (for the remaining 50%). In the case of the 4.1 measures, the national public operator has of course been playing a strategic role, by being obliged to finance the other 50% of the 4.1 measures. The important role played by national bodies in all countries has led to a different configuration of the STAR Programme in the different

[5]The EC has divided the intervention programme into two large projects. The first deals with the implementation of networks, and all interventions dealing with this project are called the '4.1 measures'. The second deals with the implementation of demonstration and promotion centres for advanced services, and all interventions regarding this project are called '4.2 measures'.

countries. A simple example of this difference is provided by comparing the Italian case with the Spanish. The logic underpinning the STAR interventions in these two countries is rather a contrasting one. In Italy, national bodies opted for a *'top-down'* intervention, i.e. the decision about the type of network or service was made at a national level, and then applied in the different areas. In Spain, in contrast, the policy chosen for the promotion of the STAR technologies was quite the opposite, i.e. a *'bottom-up'* approach, where the decisions about which networks and services had to be offered were made at the local level by the local responsible bodies. The resulting experience in terms of networks and services offered in Italy and Spain has been rather different. In Italy, the *'top-down'* approach has brought about the development of similar new networks and services in all regions of the Mezzogiorno, but with no particular interest in local needs or local requirements as expressed by local users, but with the possibility of connecting all Southern Italy with well advanced and modern infrastructures. In Spain, on the contrary, the *'bottom-up'* policy has led to the creation of networks and services customized to meet all specific local requirements and needs, but with no possibility for long distance communications.

The allocation of resources to different projects reflects the different adoption philosophies and policies, thus following different patterns in the different countries.

## 9.3 Network Externalities as the Main Reasons for Adoption: Descriptive Results

This section presents descriptive results. In particular, this incorporates a *behavioural analysis* pinpointing the most important decisions to adopt telecommunicationstechnologies. In this way, we are able to test whether our research proposition interpreting network externalities as one of the major reasons for joining the network is empirically valid.

The present section provides contingency table analyses of questions related to the adoption decision process of new networks and services; in this connection the above-mentioned behavioural analysis of corporate users forms the major focus of this part of the study. The contingency tables show the percentage of firms replying positively to each specific choice, dividing the sample between the south and the north. Correlation analyses based on Chi-Square ($X^2$) test statistics and P-values[6] are run between each reply and the regional dimension. In this way we are able to present which replies are correlated with the regional dimension.

---

[6] The Chi-Square (CHISQ) represents the degree of dependency between two variables. In our case the two variables will be the percentage of positive replies obtained in our interviews and the regional variable. The P-value represents the probability to accept the null hypothesis when this is true (McFadden, 1983).

The results on the level of adoption show higher adoption rates in the North than in the South, especially in the case of infrastructure. Among infrastructures, Packet Switching Networks and Local Area Networks (LANs) prevail, while in the services, data banks and telefax are mostly adopted.

**Table 9.1.** Adopted telecommunication networks and services by macro-areas*

| | South | North | Entire Sample | CHISQ** | P-Value*** |
|---|---|---|---|---|---|
| NETWORKS | | | | | |
| Optical Fibre Network | 2.9 | 5.7 | 4.3 | 0.35 | 0.55 |
| ISDN | 2.9 | 5.7 | 4.3 | 0.35 | 0.55 |
| Packet Switching Network | 11.4 | 31.5 | 21.4 | **0.15** | **0.04** |
| LAN | 11.4 | 40.0 | 25.7 | **7.48** | 0.00 |
| Broadband Network | 8.6 | 5.7 | 7.1 | 0.21 | 0.64 |
| SERVICES | | | | | |
| Videotex | 22.8 | 34.3 | 28.6 | 1.1 | 0.3 |
| Videoconference | 2.8 | 2.8 | 2.8 | 0.0 | 1.0 |
| Electronic Mail | 51.4 | 28.6 | 40.0 | **3.8** | **0.05** |
| Electronic Data Interchange | 17.1 | 17.1 | 17.1 | 0.0 | 1.0 |
| Databanks | 71.4 | 54.3 | 22.8 | 2.2 | 0.1 |
| Telefax | 37.2 | 100 | 68.6 | **32.1** | **0.0** |

* These values represent the percentage of positive replies obtained for each question in the two areas

** The chi-square (CHISQ) represents the degree of dependency between the percentage of positive replies and the regional variables.
The results indicating a significant dependency are reported in bold characters

*** The P-value represents the probability to accept the null hypothesis when this is true
The P-value is acceptable up to a value of 0.05 (Loehlin 1987; SAS 1989)

Concerning the results of our *sample in the South*, adoptions are overwhelmingly concentrated in 1991 and involve services much more frequently than infrastructures (Table 9.1). Among infrastructures, Packet Switching Networks and LANs are the only ones with a significant number of adoptions. This result may be explained by the fact that Packet Switching Networks can only be used for data transmission because of their technical characteristics, and data transmission is the second means of communication among firms, after voice. Among services, data banks prevail, followed by electronic mail in both service and manufacturing sectors. This means that both interactive services (such as E-mail) and non-interactive services (such as data banks) have been well received. Ranking the importance between these two popular services, data banks are ranked more highly than E-mail. This is mirrored by actual use: despite highly individual patterns, on average data banks are used on a weekly basis, whereas E-mail is only used occasionally.

The results from the sample in *the North of Italy* are quite different, as expected. Adoptions are much more distributed over the last 5 years. Among infrastructures, LANs prevail, followed by Packet Switching Networks (Table 9.1). Their higher penetration rate can easily be attributed to two main reasons: a) these are typically data transmission networks, and transmission is a necessary aspect of these technologies; b) moreover, the technical elements associated with these networks were introduced in the market some years before the technical elements supporting other kinds of networks, i.e. ISDN or Broadband Networks. This fact has two consequences; i) the technology itself is more reliable; ii) the adoption process has had more time to take place.

Among services, data banks also prevail in the North (54.3%), followed by the videotex service (34.3%) (offering amongst others also a database system), and by E-mail (28.6%). Video conferencing is not yet diffused, and, to a lesser extent, this is also true for the EDI service (17.2%), which has a lower percentage of adopters than expected (Table 9.1).

### 9.3.1 Reasons for Adoption

Interesting results stem from the *main reasons for adoption* especially if a comparison between the two macro-areas is carried out. The most important reasons for adoption in our sample are 'the importance of the technologies for the business' and 'low costs of implementation and use'. However, once the analysis is carried out at a regional level, variations in the replies emerge.

In *the South of Italy* interesting results originate from the analysis, since they support our research proposition, that the number of already existing subscribers is the main reason for adoption. Reasons put forward for adoption stress very strongly 'the existence of promotion and demonstration centres', but recorded almost as highly are the reasons concerning 'the importance of the technology for the corporate business' and 'low usage and implementation costs' (Table 9.2).

These replies witness the difficulties of adoption processes, since:

- help from the supply side in demonstrating and promoting innovation is of crucial importance. Greater possible connectivity deriving from greater availability is not in itself a sufficient element to give rise to adoption processes;
- moreover, another main reason for adoption in the first phases of a diffusion process is 'the importance of innovation for the business'. In other words, if a linkage is shown between the technology and the business areas, there is a high rate of success for an innovation;
- also stimuli on the financial side become strategic, once the risks of failure of the adoption have to be borne by the adopters. Risks are of course higher during the first phases of adoption.

Another remarkable result is the *total lack* of any references to 'the number of connected users (in the same or other regions), suppliers, buyers and competitors'. This result stresses once more the empirical plausibility of our testable hypotheses. In the first phases of the adoption process, such as in the case of the STAR Programme in the decision to adopt, other reasons prevail rather than 'the number of already existing subscribers'. In this area, in fact, there is still too small a number of subscribers, which could not be expected to act as an attracting factor for potential subscribers.

The major reason put forward for adoption by firms located in *the North of Italy* is 'the importance of these technologies for business' (more than 65.7% of replies), followed by 'low costs of implementation' (22.9%) and 'use costs' (17.1%) and by 'the high percentage of already existing subscribers in the region' and 'in other regions' (37%) (Table 9.2). These replies confirm our expectations, since in the case of an area in an advanced stage of diffusion:

**Table 9.2.** Main reasons for adoption by macro-areas

| | South | North | Entire Sample | CHISQ* | P-Value* |
|---|---|---|---|---|---|
| High percentage of already networked subscribers in your region | 5.7 | 20.0 | 12.8 | 3.18 | 0.074 |
| High percentage of already networked subscribers in other regions | 11.4 | 17.1 | 14.3 | 0.467 | 0.493 |
| Importance of the network or service for your business | 37.1 | 65.7 | 51.4 | **0.719** | **0.017** |
| Your suppliers were networked or were using the same service | 5.7 | 11.4 | 8.6 | 0.729 | 0.393 |
| Other firms in the same sector were connected | 8.6 | 14.3 | 11.4 | 0.565 | 0.452 |
| No other advanced communication network or service available | 11.4 | 5.7 | 8.6 | 0.729 | 0.393 |
| Low costs of implementation | 42.9 | 22.9 | 32.8 | 3.17 | 0.075 |
| Low costs of use | 45.7 | 17.1 | 31.4 | **3.17** | **0.01** |
| Increasing awareness of these technologies through demonstration centres | 40.0 | 5.7 | 22.8 | **11.66** | **0.001** |
| Image effect | 20.0 | 14.3 | 17.1 | 0.402 | 0.526 |
| Others | 31.4 | 14.3 | 22.8 | 2.92 | 0.088 |

* As in Table 9.1

- the total number of already existing subscribers is a crucial motive for previous adoption, once these technologies are of any interest to the business and their costs are reasonable;
- in comparison with the South, where the percentage of responses declaring this motive is around 17%, the number of subscribers plays a more crucial role in the decision process to adopt in the North. In fact, given the more advanced stage of the diffusion process in the North, one can explain this result by reiterating the fact that the higher the number of subscribers (after the critical

mass has been achieved), the more interesting the adoption and thus the 'bandwagon effect' occurs;

- in advanced stages of diffusion, the help from the supply side, through the establishment of demonstration and promotion centres, does not play any role in the diffusion process (only 5.7% of replies);
- in any stage of diffusion, price incentives (i.e. free of charge technologies in the South) play a significant role as mechanisms which help the adoption process.

### 9.3.2 Reasons for Non-Adoption

The importance of the number of adopters in the first phases of diffusion is also confirmed by another result obtained, i.e. *the reasons put forward by firms for NON-adoption*. The results from the all sample show that the most important reasons put forward by firms for NON-adoption are 'the importance of the network or service for corporate business' and 'the low percentage of connected subscribers, suppliers and customers'.

Regional variations are once more evident. In the South two such reasons prevail in particular: 'the low percentage of connected local users' and 'the irrelevance for corporate business'. Other significant adoption bottlenecks include 'the low percentage of non-local connected users and of connected suppliers and customers','the use and adoption costs (including time costs before efficient connection is achieved) and 'the inability to understand and/or evaluate these technologies' (Table 9.3).

**Table 9.3.** Main reasons for NON-adoption by macro-areas

| | South | North | Entire Sample | CHISQ* | P-VALUE* |
|---|---|---|---|---|---|
| Low percentage of connected subscribers | 71.4 | 14.3 | 42.8 | **23.33** | **0** |
| Low percentage of connected subscribers in other regions | 45.7 | 8.6 | 27.1 | **12.21** | **0** |
| None of your suppliers or customers use them | 57.1 | 31.4 | 44.3 | **4.69** | **0.03** |
| You do not think they may be useful for your business | 51.4 | 48.6 | 50.0 | 0.057 | 0.81 |
| You do not see their importance in your business | 28.6 | 14.3 | 21.4 | 2.12 | 0.41 |
| High costs of use | 22.9 | 37.1 | 30.0 | 1.70 | 0.19 |
| Other | 28.6 | 14.3 | 21.4 | 2.12 | 0.14 |

* As in Table 9.1

The 'bandwagon effect' in the advanced stages of diffusion processes is confirmed by the results of the North. In the case of 'the reasons for non-adoption' the difference between the North and the South is even greater than the case of 'the reasons for adoption'. While in the South the main reason for non-adoption was 'the low percentage of regional subscribers' (71.4%), in the North the main reason for non-adoption has been identified as 'the non-importance of these technologies for business' (48.6%), followed by 'the high costs of use' (37.1%) (Table 9.3). The reasons for non-adoption in the North have very little to do with very low levels of subscribers; they are linked rather to business interests and financial constraints.

### 9.3.3 Main conditions for future adoption

Table 9.4 presents the *main conditions for future adoptions* for both networks and services. Even from these results it is clear that 'a higher number of subscribers'**Table 9.4.** Main conditions for future adoption by macro-areas does not play an important role in stimulating adoption *in the North* (11.4% in general; 8.6% in the case of customers and suppliers), where the interest in future conditions for adoption is more linked to 'the technical progress of networks' (20%), to 'an increase in their efficiency' (14.3% and 'a reduction in the price of use for services' (11.4%). On the contrary, in the South the main condition for future adoption is related to 'a higher number of subscribers', for networks (28.6%) and services (40%) (Table 9.4). For services, the results are different from the case of networks. While 'the reduction in the price of use' still remains rather crucial in the North (20%), this has the same weight as 'a higher number of suppliers and customers connected' (20%). 'Technical progress' and 'increase in the efficiency of the service' both still play a crucial role (17.1%). *In the South*, instead 'a higher number of people connected' is stated to be the most important reason, especially in the case of advanced interactive services, such as videotex and E-mail. For data banks, too, 'a higher number of people connected' is the main expected condition for future adoption, explained by the fact that a higher number of people connected would assure a higher revenue to the service managers, and thus stimulate a better quality of the service (i.e. a larger variety of available information). For E-mail and Electronic Data Interchange services the most important reasons stimulating future adoption are also 'technical progress' and 'increase in network efficiency'.

### 9.3.4 Reasons for dissatisfaction with the telecommunications technologies in use

The most interesting result is the strong regional variation in the *reasons for the dissatisfaction with the quality of existing technologies.* As expected, 'the degree of satisfaction with the technologies' among subscribers is higher in the North than in the South (68.6% against 57.1%) (Table 9.5). Moreover, the main reason put forward for a high degree of dissatisfaction is 'the too low number of subscribers' (46% and 'the too high costs of implementation" (30%).

**Table 9.4.** Main conditions for future adoption by macro-areas

| | South | North | Entire Sample | CHISQ* | P-VALUE* |
|---|---|---|---|---|---|
| NETWORKS | | | | | |
| A higher number of people connected | 28.6 | 11.4 | 20.0 | 3.14 | 0.07 |
| A higher number of suppliers and customers connected | 25.7 | 8.6 | 17.1 | **3.62** | **0.06** |
| A better geographic distribution of the network | 11.4 | 5.7 | 8.6 | 0.73 | 0.39 |
| A reduction in the price of use | 11.4 | 11.4 | 11.4 | 0.0 | 1 |
| A reduction in the price of access | 8.6 | 8.6 | 8.6 | 0.0 | 1 |
| More recent availability of the network | 8.6 | 11.4 | 10.0 | 0.16 | 0.69 |
| Technical progress in the network | 25.7 | 20.0 | 22.8 | 0.32 | 0.57 |
| Increase in the efficiency of the network | 20.0 | 14.3 | 17.1 | 0.40 | 0.53 |
| Good results from the previous adoption | 14.2 | 5.7 | 10.0 | 1.43 | 0.23 |
| SERVICES | | | | | |
| A higher number of people connected | 40.0 | 14.3 | 27.1 | **5.85** | **0.02** |
| A higher number of suppliers and customers connected | 31.4 | 20.0 | 25.7 | 1.19 | 0.27 |
| A better geographic distribution of the service | 17.1 | 2.9 | 10.0 | **3.97** | **0.05** |
| A reduction the price of use | 5.7 | 20.0 | 12.9 | 3.2 | 0.07 |
| A reduction in the price of access | 11.4 | 17.1 | 14.3 | 0.47 | 0.49 |
| More recent availability of the service | 11.4 | 5.7 | 8.6 | 0.73 | 0.39 |
| Technical progress in the service | 20.0 | 17.1 | 18.6 | 0.09 | 0.76 |
| Increase in the efficient of the service | 25.7 | 17.1 | 21.4 | 0.76 | 0.38 |
| Good results from the previous adoption | 20.0 | 2.9 | 11.4 | **5.08** | **0.02** |

* As in Table 9.1

**Table 9.5.** Degree of satisfaction of the quality of communication by macro-areas

| | South | North | Entire Sample | CHISQ* | P-VALUE* |
|---|---|---|---|---|---|
| YES | 57.1 | 68.6 | 62.8 | 0.98 | 0.322 |
| NO | 42.9 | 31.4 | 37.1 | | |

* As in Table 9.1

It is interesting to see that these two reasons are also very much dependent on the regional dimension. While in the North 'the high costs of their implementation' are the most important reasons (54.5%) this does not apply to the South (13.3%) (Table 9.6). Dissatisfaction does not stem from 'the low number of subscribers' (only 9.1% of positive replies for this reason), as is the case in the South (73.3%).

**Table 9.6.** Major persistent user problems, in the case of dissatisfaction by micro-areas

| | South | North | Entire Sample | CHISQ* | P-Value* |
|---|---|---|---|---|---|
| Too high costs of implementation | 13.3 | 54.5 | 30.7 | **5.06** | **0.024** |
| They are no longer free charge services | 20.0 | 0.0 | 11.5 | 2.48 | 0.11 |
| The use of these technologies requires requires profound organisational | 0.0 | 9.1 | 3.8 | 1.42 | 0.23 |
| The number of subscribers using these services and networks is still too low | 73.3 | 9.1 | 46.1 | **10.53** | **0.001** |
| Suppliers are not using these technologies | 33.3 | 9.1 | 23.1 | 2.10 | 0.15 |
| Competitors are not using these technologies | 0.0 | 26.7 | 15.4 | 3.47 | 0.063 |
| Other | 54.5 | 20.0 | 34.6 | 3.35 | 0.067 |

* As in Table 9.1

The sample in the North generally shows a greater interest in these technologies, witnessed by a high percentage of replies recording 'the increased intensity of business relationships after the adoption of telecommunications technologies' (62.9% of replies in the North against 20% in the South (Table 9.7). For the 37% of 'decreased business relationships after the introduction of these technologies' are put forward as the main reasons for constant business relationships. Again, these results witness the fact that at low penetration levels, such as in the South, 'the low number of subscribers' becomes a constraint for the increase in the intensity of use of these technologies. In the case of more advanced adoption processes, the reason explaining the low level of intensity of use is 'the non-importance for business', represented by 14.3% of replies.

Table 9.7. Intensity of relationships after telecommunication technology adoption by macro-areas

| | South | North | Entire Sample | CHISQ* | P-VALUE* |
|---|---|---|---|---|---|
| Increased intensity of business relationships | 20.0 | 62.9 | 41.4 | **13.25** | **0.0** |
| Constant intensity of business relationships | 0 | 0 | 0 | | |
| Decreased intensity of business relationships | 80.0 | 37.1 | 58.6 | | |

* As in Table 9.1

The results presented are quite satisfactory, since at least at a descriptive level, our expectations are fulfilled and our testable hypotheses verified with *strong regional variations,* as expected. The importance of the regional dimension in our replies is witnessed by the statistically significant level of $\chi^2$ and of the P-value between the 'network externality' variables and the regional dimension only for the network externality variables, as is represented by the contingency tables, where the statistically significant dependency is marked by bold print (Tables 9.2/9.8).

These results reinforce our expectations based on the general beliefs in the theoretical literature on consumption network externalities that in the first phases of adoption of new networks and services (such as in the South), the main reason for non-adoption is 'the low percentage of subscribers already linked to the network'. It makes sense, therefore, to find that the main reasons for adoption in the very first phases of the adoption process is *not* the number of already existing subscribers. Nevertheless, one of the major problems still existing in the use of new telecommunications technologies is the low number of subscribers. It should be noted, however, that all these results do not hold in the case of Northern Italy, where more advanced diffusion processes lead to opposite results, as expected.

## 9.4 Conclusions[7]

In this paper we have focused on a behavioural analysis of the main reasons for adoption of advanced telecommunications technologies. In particular, we have presented a descriptive analysis, through contingency tables analysis.

---

[7]The first author greatly acknowledges funds of the Italian National Research Council (CNR) - project n. 94.00560.CTH, directed by Prof. Roberto Camagni. Though the study is the result of a joint effort of the two authors, R. Capello wrote Sections 14.2 and 14.3. The remaining Sections have been jointly written.

The results strengthen the role of the number of already existing subscribers as one of the main reasons for adoption and emphasize the *regional differences* in the reasons for joining the network. In backward regions, where a critical mass has not yet been achieved, the reasons for adoption are not the number of already existing subscribers. On the contrary, in advanced regions, the number of adopters represents one of the most important reasons for adoption. As expected, the results are the opposite when dealing with the main conditions for future adoption. In backward regions, the most important reasons for future adoption lie in an increase in the number of subscribers connected, while for advanced regions other reasons, such as low implementation and use costs, are relevant for future adoption. Moreover, an interesting result obtained from the contingency tables analysis is the statistical dependency of 'consumption network externality variables' with the regional dimension. This result is even more important when one remarks that consumption network externality variables are *the only ones* showing a statistical dependency with the regional dimension.

**Table 9.8.** Main reasons in case of unincreased itensity by macro areas

| | South | North | Entire Sample | CHISQ* | P-VALUE* |
|---|---|---|---|---|---|
| None | 20.0 | 11.4 | 15.7 | 0.97 | 0.32 |
| Costs of initial investment | 2.8 | 0.0 | 1.4 | 1.01 | 0.32 |
| Lack of reliability on the technologies used | 2.8 | 0.0 | 1.4 | 1.01 | 0.34 |
| Level of telecommunication charges | 0.0 | 0.0 | 0.0 | | |
| Lack of staff skills in their use | 2.8 | 14.3 | 8.6 | 2.92 | 0.08 |
| Services are not relevant to business | 14.3 | 17.1 | 15.7 | 0.11 | 0.74 |
| Your customers do not use them | 0.0 | 31.4 | 15.7 | **13.05** | **0.0** |
| Your suppliers do not use them | 2.8 | 25.7 | 14.3 | **7.45** | **0.006** |
| Too few subcribers in general | 40.0 | 0.0 | 20.0 | **17.5** | **0.0** |
| Restricted geographical diffusion of networks and services | 8.6 | 2.8 | 5.71 | 1.06 | 0.30 |
| Other reasons | 8.6 | 5.7 | 7.14 | 0.21 | 0.64 |

* As in Table 9.1

As far as the STAR programme is concerned, our analysis clearly demonstrates the importance the number of subscribers has in the first stages of diffusion. However, to evaluate the degree of success of the Programme, it is necessary to test its effects on the performance of (adopting) firms and regions. In other words, whether *production network externalities are exploited by firms and regions* must be tested empirically. This is subject matter that the two authors are developing.

It is our intention to try and see whether the positive and interesting results obtained with this descriptive analysis are confirmed with an interpretative analysis.

If the results are confirmed, some interesting and strategic policy guidelines emerge in the future for policy makers. A broader analysis is in fact able to show which are the most important mechanisms for the diffusion of these technologies and which are the strong points on which a 'possible STAR 2' programme may focus its attention more, in order to get best results in terms of adoption and, consequently, of positive economic effects on regional development.

## References


Allen D (1988) New telecommunications services. Network externalities and critical mass. Telecommunications Policy, September 257-271

Allen D (1989) Competition, cooperation and critical mass in the evolution of networks. Paper presented at the 8th ITS Conference. 18-24 March, Venice

Antonelli C (1989) The diffusion of information technology and the demand for telecommunications services. Telecommunications Policy, September 255-264

Antonelli C (1990) Induced adoption and externalities in the regional diffusion of information technology. Regional Studies 24:31-40

Antonelli C (ed) (1992) The Economics of Information Networks. Elsevier Science Publisher

Bental B, Spiegel M (1990) Consumption externalities in telecommunication services. In: de Fontenay M, Sibley D (Eds) Telecommunications Demand Modelling. Elsevier Science Publisher

Cabral L, Leite A (1992) Network consumption externalities. The case of portuguese telex service. In: Antonelli C (Ed) 129-140

Capello R (1994) Spatial Economic Analysis of Telecommunications Network Externalities. Avebury, Aldershot

Capello R, Nijkamp P (1994) Regional Variations in Production Network externalities. Paper presented at NECTAR, January, Vienna, 28-29

Cowen T (ed) (1988) The Theory of Market Failure. A Critical Examination. George Mason University Press

Curien N, Gensollen M (1987) A functional analysis of the network. A prerequisite for deregulating the telecommunications industry. Annales des télécommunications, 42:629-641

David P (1985) Clio and the economics of Qwerty. AEA Papers and Proceedings, 75:332-337

David P (1992) Information network economics. Externalities, innovation and evolution. In: Antonelli C (Ed) 103-106

Farrell J, Saloner G (1985) Standardisation, compatibility, and innovation. Rand Journal of Economics, 16:70-83

Hayashi K (1992) From network externalities to interconnection: the changing nature of networks and economy. In: Antonelli C (Ed) 195-216
Katz M, Shapiro C (1986) Technology adoption in the presence of network externalities. Journal of Political Economy, 822-841
Markus M (1989) Critical mass contingencies for telecommunications consumers. Paper presented at the 8th ITS Conference, 18-24 March, Venice
Mishan EJ (1971) The Postwar Literature on Externalities. An Interpretative Essay. The Journal of Economic Literature, 9:1-28
Rohlfs J (1974) A theory of interdependent demand for a communication service. The Bell Journal of Economics and Management Science 5:16-37

# 10 Connectivity and Congestion on European Road Corridors

A. Reggiani, P. Nijkamp and G. Pepping

## 10.1 Introduction

Europe is increasingly moving towards a network economy, in which physical flows of people and goods form the connecting elements. This means that European corridors - especially those between major metropolitan areas or centres of economic activity in Europe - are playing a critical role in the new, evolving map of Europe. At the same time it has to be recognized that traffic congestion on many European transport corridors is dramatically increasing, while also the environmental implications and the poor safety conditions cause unacceptable social costs. Conventional transport policy is apparently unable to keep up with Europe's growing demand for mobility. New transport technologies are hence increasingly regarded as necessary instruments for ensuring sustainable mobility.

Advanced Transport Telematics (ATT) is nowadays generally regarded as a potentially powerful tool for coping with future transport problems on European corridors.

It is clear, however, that not all European corridors present the same characteristics. Therefore, it is necessary to develop an appropriate typology of European corridors on the basis of the indigenous relevant features of these corridors and also of the potential of relevant ATT applications. The rationale behind the identification of European corridor typologies is the idea that European policy resolutions (like ATT introduction) may generate different behavioural responses for each type of corridor and ATT technology.

In the context of the present paper, particular attention will be given to the role of peripheral corridors (such as the 'Mediterranean' corridors connecting Greece).

## 10.2 Description of the Corridors Investigated

The identificaton of relevant types of road corridors is not an easy task. The first step of the present study is the selection of relevant inter-urban corridors, i.e. the identification of those inter-urban corridors whose characteristics may be considered relevant for studying the various potential impacts of ATT. In particular, this can be undertaken by investigating two aspects of the problem, viz. the expected ATT impacts (on system's efficiency, on network reliability, on environmental quality, etc.), and the different actors interacting within a comprehensive transport system (i.e., users, operators, public decision-makers and decision-makers in governments and the political system as well as in the economic and social system).

These interacting factors lead to the following two categories of criteria that are useful for the selection of the corridors:

- the *spatial location* of the corridor, which refers both to its geographical coverage (for instance, the corridors considered should contain not only central, but also central-peripheral and peripheral patterns of interaction) and the strategic relevance of the corridor (in particular, the strategic importance of the corridor in the European communication network and also in the increasing process of regional integration);
- the *different mobility patterns* within the regions crossed by the corridors (especially various aspects of traffic flows, such as degrees of congestion, pollution, network reliability, safety, etc.) in combination with different socio-economic situations (in terms of wealth, productive specialization, etc.).

It also has to be added that quite some field work on various corridors has already been undertaken, so that it makes sense to focus in particular on DRIVE 2 projects undertaken by the EC. Therefore, in addition to four pilot projects in DRIVE (PLEIADES, META-FORA, MELYSSA, ADS), three other DRIVE 2 projects will be considered (INVAID, COMBICOM, FRAME) which all together cover Centre-East and Centre-West directions. However, in order to deal with a more balanced set of European mobility patterns, an additional five corridors are taken into account reflecting East-West directions (such as the Trieste-Lisbon and Vienna-Nantes corridors) as well as inner links in the core of Europe (such as London-Berlin, Amsterdam-Milan and Amsterdam-Utrecht) (see Figure 10.1).

Consequently, on the basis of the above criteria the following list of corridors has been identified:

1. Glasgow-London-Paris-Marseille (PLEIADES-DRIVE)
2. London-Athens (METAFORA-DRIVE)
3. Stuttgart-Lyon-Barcelona (MELYSSA-DRIVE)
4. Aosta-Athens (ADS-DRIVE)
5. Amsterdam-Madrid (INVAID 2-DRIVE)
6. Rotterdam-Munchen-Verona (COMBICOM-DRIVE)
7. Amsterdam-Athens (FRAME-DRIVE)
8. Trieste-Lisbon
9. Wien-Nantes
10. London-Berlin
11. Amsterdam-Milan
12. Amsterdam-Utrecht

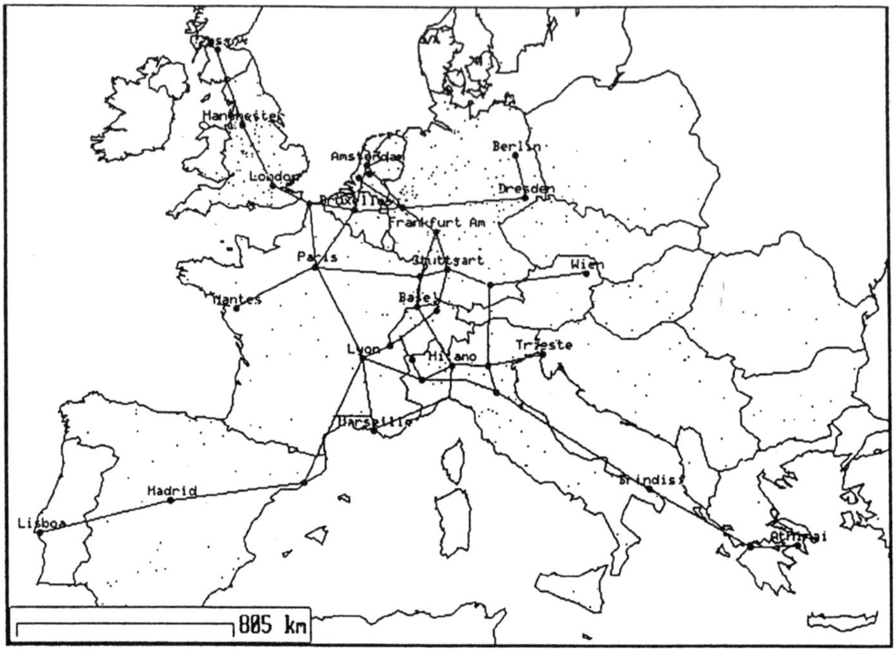


**Fig. 10.1.** Relevant European inter-urban corridors

Each corridor route is identified on the basis of the main road infrastructure (first motorway, otherwise highway; see Figure 10.1). Figure 10.1 also shows that corridors can be subdivided into shorter segments connecting two relevant nodes along the route.

Based on the analysis of the corridor segments and the related nodes within corridors, it is also possible to make some observations on regional interactions and relevant nodes on the network (see Table 10.2). Firstly, three important 'hubs' in the regional interactions emerge: the Baden-Wurttemberg region (around Stuttgart) expanding towards the North (Nordrhein-Westphalia and West-Netherlands) and the South (Bavaria and North Italy), the Nord-Pas-de-Calais region linking England to the European continent, and the French Centre-East region interacting with the Paris region, Spain and the North of Italy.

As regards the nodes (36) identified on the network, Stuttgart and Frankfurt are those with the greatest number of inter-urban links (they are crossed by 6 and 5 corridors respectively), confirming the hypothesis that the German axis is one of most strategic axes in the European mobility system, both in the centre/centre-periphery interaction and in a North-South/East-West direction.

These two nodes are followed by a series of nodes which are crossed by four corridors. They are: Amsterdam (which mainly emerges as a terminal point of three important inter-urban corridors, viz. Amsterdam-Madrid in North/South-West direction, Amsterdam-Milan in North/South direction and

Amsterdam-Athens in North/South-East direction), Cologne and Munich (which form the strongest axis in Germany) and Verona and Lyon, (whose main characteristic is to act as an important crossroad for North/South and East/West movements).

By considering the nodes which are crossed by up to three corridors, it is possible to gain strategic insights into the relevant axes in Europe. In fact, three other main inter-urban links emerge, notably the triangle London-Brussels-Paris, the Spanish axis between Barcelona and Madrid, and the Italian Adriatic axis to Greece (in particular, the corridor 'Aosta-Athens'; see Figure 10.2). Finally, the nodes of Rotterdam, Strasbourg, Milan and Torino serve a number of corridors at a minor level, despite their seemingly central position. It is interesting to note that the above classification of nodes also reflects the so-called European pentagon model conceived by Nakicenovic (1992), which regards as the core of Europe the socio-economic and traffic network among the five centres of Amsterdam, London, Paris, Milan and Berlin.

This qualitative analysis of corridors will be the basis for a more detailed investigation of the main differences and similarities of the corridors in order to construct a European typology in light of potential ATT applications (see Sections 10.4 and 10.5). A further examination and interpretation of the related data will be the subject of the next section.

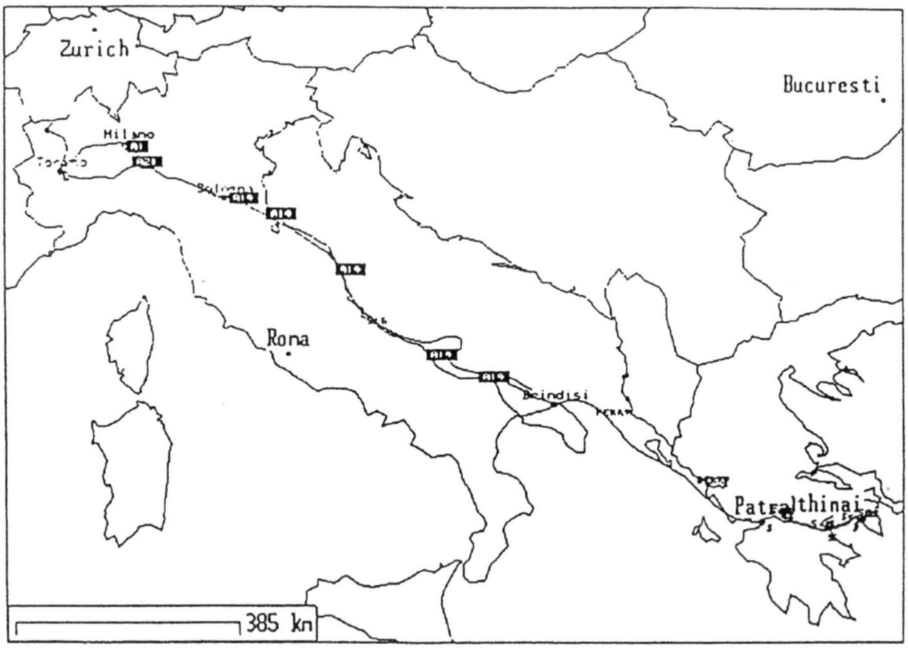


**Fig. 10.2.** The corridor Aosta-Athens

**Table 10.1.** Corridor basic attributes

| Corridors | Length | Relevant Nodes | Crossed Regions(*) |
|---|---|---|---|
| Amsterdam-Utrecht | 40 | Amsterdam, Utrecht | NL3 |
| Amsterdam-Milan | 1200 | Amsterdam, Koln, Frankfurt, Stuttgart, Strasbourg, Basel, Milan | NL3,NL4,GE5,GE6 GE8,SW1,IT2 |
| London-Berlin | 1300 | London, Calais Brussels, Frankfurt, | UK5,FR3,BE1,BE2 BE3,GE5,GE6 |
| Rotterdam-Verona | 1200 | Rotterdam, Koln, Frankfurt, Stuttgart, Munich, Verona | NL3,NL2,GE5,GE6 GE8,GE9,AUS,IT3 |
| London-Athens | 2350 | London, Calais, Brussels, Koln, Frankfurt, Stuttgart Munich, Verona, Bologna, Brindisi, Patra Athens | UK5,FR3,BE1.BE2 GE5,GE6,GE8,GE9 AUS,SWi,IT3,IT4 IT5,IT8,IT9 GR2 |
| Amsterdam-Athens | 2300 | Amsterdam, Koln, Frankfurt, Stuttgart, Munich, Verona, Bologna, Brindisi, Patra, Athens | NL2,NL3,GE5,GE6 GE8,GE9,AUS,SWI IT3,IT4,IT5,IT8 IT9,GR2 |
| Glasgow-Marseille | 1700 | Glasgow, Manchester London, Calais, Paris Lyon, Marseille | UK10,UK1,UK8,UK7 UK5,FR3,FR1,FR2 FR7,FR8 |
| Amsterdam-Madrid | 2250 | Amsterdam, Rotterdam Brussels, Paris, Lyon Barcelona, Madrid | NL3,NL2,BE1,BE3 FR7,FR8,SP5,SP2 SP4,SP3 |
| Aosta-Athens | 1350 | Aosta, Torino, Bologna, Brindisi, Patra, Athens | IT2,IT4,IT5,IT8 IT9,GR2 |
| Vienna-Nantes | 1600 | Vienna, Munich, Stuttgart, Strasbourg, Paris, Nantes | FR5,FR2,FR1,FR4 GE8,GE9 |
| Trieste-Lisbon | 2800 | Trieste, Venice, Verona Milan, Torino, Lyon, Barcelon, Madrid, Lisbon | PO2,SP3,SP4,SP2 SP5,FR8,FR7,IT1 IT2,IT3 |
| Stuttgart-Madrid | 1900 | Stuttgart, Zurich, Geneva, Lyon Barcelona, Madrid | GE8,AUS,FR7 FR8,SP5,SP2,SP4 SP3 |

Legend:

| | | | |
|---|---|---|---|
| AUS Austrie | GE5 Nordrhein-Westfale | IT7 Campania | SP5 Este |
| BE1 Vlaams Gewest | GE6 Hessen | IT8 Abruzzi-Molise | SP6 Sur |
| BE2 Region Wallonne | GE7 Rheinland-Pfalz | IT9 Sud | SP7 Canarias |
| BE3 Bruxelles | GE8 Baden-Wurttemberg | IT10 Sicilia | SWI Switzerland |
| DK Danimarca | GE9 Bayern | IT11 Sardegna | UK1 North |
| FR1 Ile-de-France | GE10 Saarland | L Lussemburgo (GD) | UK2 Yorkshire-Humbers |
| FR2 Bassin Parisien | GE11 Berlin West | NL1 Noord Nederland | UK3 East Midlands |
| FR3 Nord-Pas-de-Calais | GR1 Voreia | NL2 Oost Nederland | UK4 East Anglia |
| FR4 Est | GR2 Kentrike | NL3 West Nederland | UK5 South-East |
| FR5 Ouest | GR3 Anatolia E Nord | NL4 Zuid Nederland | UK6 South-West |
| FR6 Sud-Ouest | IR Irlanda | P01 Nord | UK7 West Midlands |
| FR7 Centre-Est | IT1 Nord-Ovest | PO2 Sud | UK8 North-West |
| FR8 Mediterranee | IT2 Lombardia | P03 Isole | UK9 Wales |
| GE1 Schleswig-Holstein | IT3 Nord-Est | SP1 Noroeste | UK10 Scotland |
| GE2 Hambourg | IT4 Emilia-Romagna | SP2 Noreste | UK11 Northern Ireland |
| GE3 Niedersachsen | IT5 Centro | SP3 Madrid | |
| GE4 Bremen | IT6 Lazio | SP4 Centro | |

* The regional disaggregations is based on the nomenclature of territorial units for statics (NUTS I level according to the Eurostat regional statistics)

**Table 10.2.** Frequency analysis for the nodes

| Nodes | Number of Corridors | Corridors Identified (*) |
|---|---|---|
| Stuttgart | 6 | AMS-MIL; ROT-VER, LON-ATH- AMS-ATH, VIE-NAN- STU-MAD |
| Frankfurt | 5 | AMS-MIL; LON-BER; ROT-VER, LON-ATH, AMS-ATH |
| Amsterdam | 4 | AMS-UTR, AMS-MIL, AMS-ATH; AMS-MAD |
| Koln | 4 | AMS-MIL- ROT-VER- LON-ATH- AMS-ATH |
| Lyon | 4 | GLA-MAR- AMS-MAD- TRI-LIS, STU-MAD |
| Munich | 4 | ROT-VER, LON-ATH; AMS-ATH; VIE-NAN |
| Verona | 4 | ROT-VER, LON-ATH; AMS-ATH; TRI-LIS |
| Brussels | 3 | LON-ATH; LON-BER; AMS-MAD |
| Calais | 3 | LON-BER- LON-ATH- GLA-MAR |
| London | 3 | LON-BER, LON-ATH, GLA-MAR |
| Paris | 3 | GLA-MAR; VIE-NAN- AMS-MAD |
| Barcelona | 3 | AMS-MAD, TRI-LIS, STU-MAD |
| Madrid | 3 | AMS-MAD- TRI-LIS- STU-MAD |
| Bologna | 3 | LON-ATH; AMS-ATH; AOS-ATH |
| Brindisi | 3 | LON-ATH; AMS-ATH; AOS-ATH |
| Patra | 3 | LON-ATH; AMS-ATH; AOS-ATH |
| Athens | 3 | LON-ATH; AMS-ATH; AOS-ATH |
| Rotterdam | 2 | ROT-VER- AMS-MAD |
| Strasbourg | 2 | AMS-MIL; VIE-NAN |
| Milan | 2 | AMS-MIL- AOS-ATH |
| Torino | 2 | AOS-ATH, TRI-LIS |
| Aosta | I | AOS-ATH |
| Basel | I | AMS-MIL |
| Berlin | I | LON-BER |
| Dresden | I | LON-BER |
| Geneve | I | STU-MAD |
| Glasgow | I | GLA-MAR |
| Lisbon | I | TRI-LIS |
| Manchester | I | GLA-MAR |
| Marseille | I | GLA-MAR |
| Nantes | I | VIE-NAN |
| Trieste | I | TRI-LIS |
| Utrecht | I | AMS-UTR |
| Venice | I | TRI-LIS |
| Vienna | I | VIE-NAN |
| Zurich | I | STU-MAD |

(*)In the sequel, the first three letters of the names of the terminal nodes are used as a code for each corridor

## 10.3 Selection of Corridor Attributes: Data Base and Interpretation

### 10.3.1 The Selection of Attributes

In order to study the differences in the potential ATT impact as a result of different corridor contexts (defined by a given corridor typology), a first step is the construction of the 'states' of the corridors, i.e. the main categories of criteria necessary for identifying the characteristics of an inter-urban corridor consistent with the objective of assessing the ATT potential. Thus, these 'states' of the corridors are characterized by relevant variables (or corridor features) in light of a possible ATT introduction. In particular, we have taken into account the threefold ATT potential effect on:

- the mobility system's efficiency, particularly with regard to lower management costs and use by operators, and to new network services and capacities, etc.;
- network reliability, with regard to better safety and information and service levels for users;
- environmental quality, particularly with regard to pollution reduction.

From these features, a (preliminary) list of interesting 'states' of the corridors can be derived. These can be expressed in a qualitative sense as follows:

- *socio-economic conditions* (within the regions crossed) which refer to different mobility patterns and, as a consequence, to different needs in terms of ATT functions as well as to different behavioural responses;
- *geographical position* (in other words, the comparative locational advantage or the strategic network position of the corridor) which is useful to capture its strategic importance[1] and its effective function;
- *corridor management*, which deals with the organizational/managerial perspective of the network corridor at hand. It tries to mirror the level of the infrastructure supply as well as the degree of complexity of such management. This concept indicates that the different degrees of interrelation/cooperation among transport operators along the corridors, the different complexities of infrastructural management, the different numbers of managers in infrastructural administration, the different operative strategies adopted and, finally, the different levels of technological innovation already achieved, do influence the ATT impact on that particular corridor;
- *traffic conditions*, with particular reference to (i) the degree of congestion (i.e., the identification of bottlenecks along the corridors generated by an insufficient supply of infrastructure or by an excess of demand) including its impact on corridor management efficiency and on user costs; (ii) safety (reliability or accident rates on the corridor), which will impact on the social (direct or indirect) benefits linked to the introduction of ATT;

---

[1]The strategic network position may be described in relation to the various functions offered by the corridor, like internal integration, interregional exchange or regional crossing (transit).

- *environmental quality* (acoustic or air pollution), which appears to be one of the principal incentives for implementing ATT.

Tentative assessments of the association between 'states' of the corridors and the potential impacts of ATT are given in Table 10.3 (for details on the various kinds of potential impacts of ATT see Nijkamp et al, 1993).

On the basis of the above considerations, appropriate data will be selected on specific indicators to be outlined in the next subsection.

**Table 10.3.** Exploratory relationships between the state of the corridors and the potential impacts of ATT.

| socio-econ. conditions | | | geographical conditions | | | corridor manageme-ment | | | traffic conditions: congestion | | | traffic condition: safety | | | environm. quality | | |
|---|---|---|---|---|---|---|---|---|---|---|---|---|---|---|---|---|---|
| LO | ME | HI | C | SP | P | LO | ME | HI | LO | ME | HI | LO | ME | HI | LO | ME | I |
| +++ | ++ | + | + | ++ | +++ | 0 | +++ | + | 0 | +++ | + | +++ | ++ | + | +++ | ++ | + |

Legend:

LO, ME, HI = low, medium/high and very high performance levels of corridors, respectively
C, SP, P = central, semi-peripheral and peripheral locational conditions, respectively
0 = negligible impact
+ ++ +++ = increasing impact levels (in terms of positive performance)

### 10.3.2 The Data Matrix

In this subsection, the corridor characteristics needed for the construction of an inter-urban corridor typology will be presented. In total, 19 different features are considered. As mentioned before, they may be regarded as factors influencing potential impacts of ATT in terms of user benefits, efficiency, safety and environmental quality. The classification of these features is based on the five categories discussed in Subsection 10.3.1.

This leads to a series of variables whose characteristics (source, year and unit of measurement) are contained in Table 10.4.

Most of the indicators used concern measurable variables. Many of them are taken from Eurostat Regional Statistics, in which extended definitions can be found. Details on the specific data can be found in Nijkamp et al, (1993) and Reggiani et al (1994).

## 10.4 Comparison of Corridors: Common Elements and Differences Relative to Potential ATT Applications

As indicated in the previous sections, the selected attributes of corridors contain relevant information on the potential impact of ATT applications. In this section the main differences/similarities will be examined in order to show the main characteristics of the selected inter-urban corridors and to identify preliminary categories and classifications of such corridors.

**Table 10.4.** List of features for the corridor typology with source, year and unit of measurement

| Variable | Source | Year | Unit of Measurement |
|---|---|---|---|
| *Socio-economic* | | | |
| 1 population density | Eurostat | 1985 | habitants per km2 |
| 2 (Total) GVAMP (*) | Eurostat | 1984 | ECU per habitant |
| 3 GVAMP of industrial sectors | Eurostat | 1984 | ECU per habitant |
| 4 GVAMP of market services | Eurostat | 1984 | ECU per habitant |
| 5 employment | Eurostat | 1984 | employment places per km2 |
| 6 employment in industrial branches | Eurostat | 1984 | employment places per km2 |
| 7 employment in market service branches | Eurostat | 1984 | employment places per km2 |
| 8 stock of dwellings | Eurostat | 1985 | dwellings per km2 |
| *Locational* | | | |
| 9 large agglomerations on corridor | - | 1992 | number of large agglomerations on corridor |
| 10 crossed EC member states | - | 1992 | number of national borders crossed by corridor |
| *Infrastructural* | | | |
| 11 motorway density | Eurostat | 1985 | length of motorway(km)per km2 |
| 12 accessibility of crossed | Keeble | 1982 | accessibility index, see explanation zones |
| *Traffic demand* | | | |
| 13 private car ownership | Eurostat | 1985 | private owned cars per km2 |
| 14 road traffic accidents | Eurostat | 1985 | motor vehicle traffic accidents per km2 |
| 15 vehicle flows | NEA | 1985 | passing vehicles in yearly daily averages |
| 16 growth of vehicle flows | NEA | 1985 | relative growth of vehicle flows between 1980 and 1985 |
| 17 heavy vehicle flows | NEA | 1985 | share of heavy vehicles |
| 18 traffic flow smoothness | NEA | 1985 | level of service, see explanation |
| *Environmental* | | | |
| 19 NOx pollutions | Europe 2000 (CEC) | 1985 per | NOx pollution in tons per km2 year |

(*) GVAMP: gross value added at market prices.

From the point of view of *'socio-economic' attributes* (see Figures 10.3 and 10.4), a subdivision between central (C) and peripheral (P) European regions immediately emerges. In particular, it can be noted that by shifting from the European peripheral regions towards the central ones (in other words, moving from the corridors connecting the centre-periphery towards the corridors joining central nodes in Europe), the residential density increases as well as the service share in total employment.

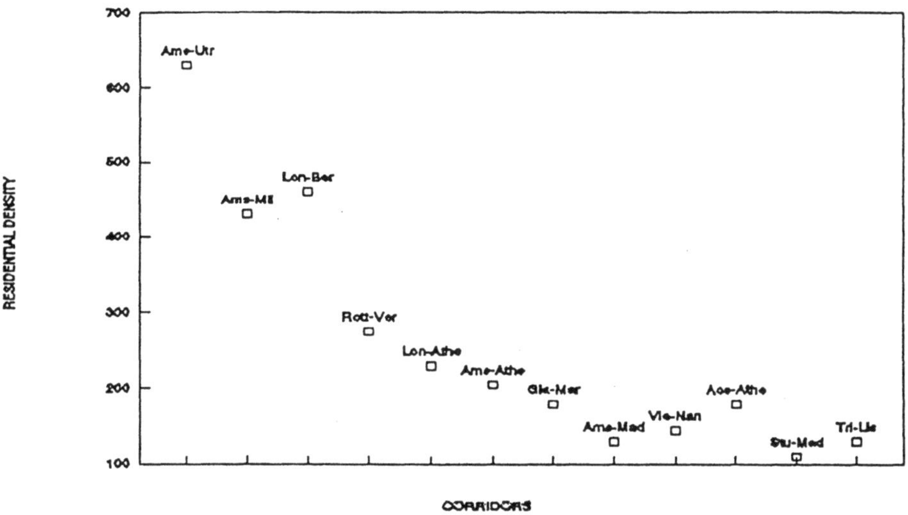


**Fig. 10.3.** Residential density (1987) (Habitants per km2)

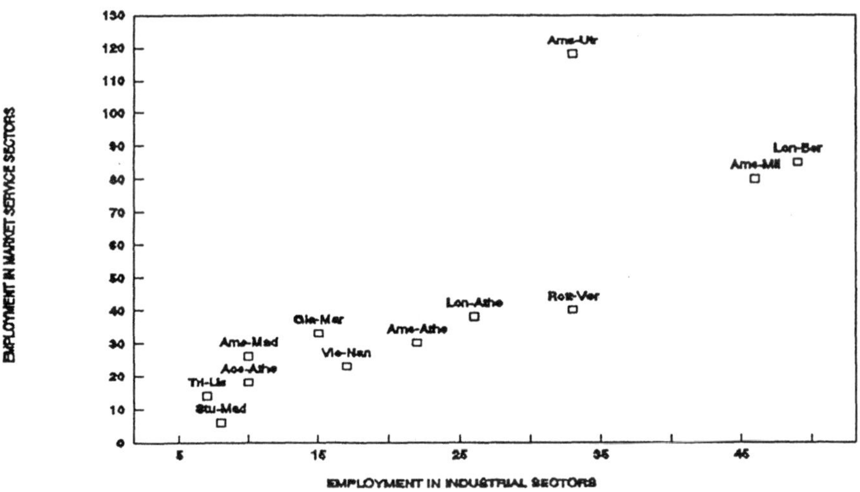


**Fig. 10.4.** Industrial products versus market service employment density (1987)

In addition, by considering only the centre-periphery corridors, differences in the residential density appear, since those in the North-South direction have an intermediate level of residential density with respect to those in the East-West direction (which show a low value).

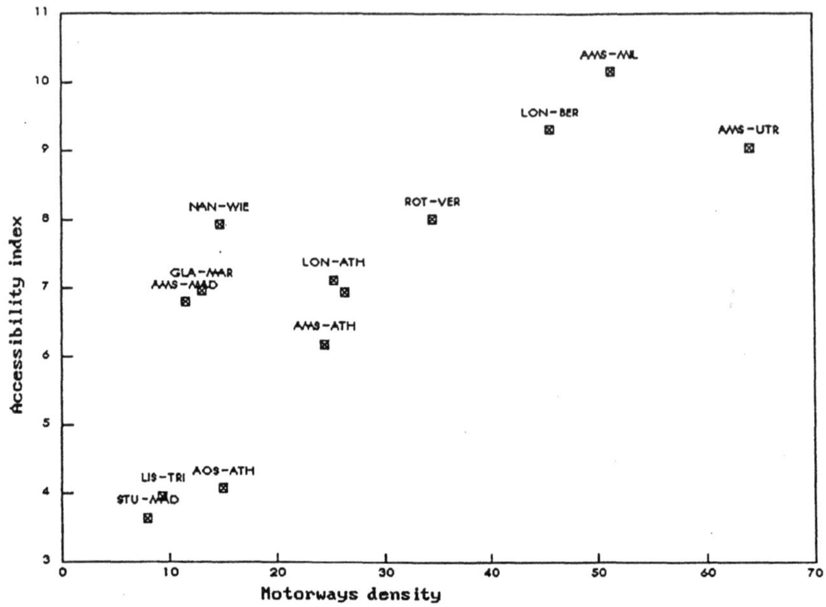


**Fig. 10.5.** Motorway density versus accessibility index

As a consequence, from these different combinations of socio-economic attributes *different needs,* both in terms of mobility and of specific ATT implementations, as well as *different reactions* of potential users are likely to emerge. The differences in mobility (in terms of both the functioning of the system and the user needs) are confirmed by considering the *infrastructure supply side* (see Figure 10.5). In particular, Figure 10.5 shows, on the basis of the motorway density, the emergence of a two-fold classification, viz. central corridors and peripheral corridors, while on the basis of an indicator of accessibility (which reflects the different effects of dominance and dependence among the regions) a further subdivision between quasi-peripheral and peripheral corridors appears. This is mirrored by a high, medium and low supply of infrastructure.

By combining *transport supply and socio-economic attributes,* a *classification* emerges based mainly on the *geographical position* of the corridors with respect to the 'core' of Europe.

With reference to the *transport demand attributes,* it is possible to identify the same type of classification which mainly reflects the different congestion situations. In particular, by investigating Figures 10.6 and 10.7 we can see that the peripheral corridors are characterized by a low accessibility index as well as a low value of traffic flows. Next, by considering the level of services related to traffic flows and the share of freight transport on total flows (see Figures 10.8 and 10.9, respectively) again the peripheral corridors appear to be clustered and clearly separated from the other central corridors.

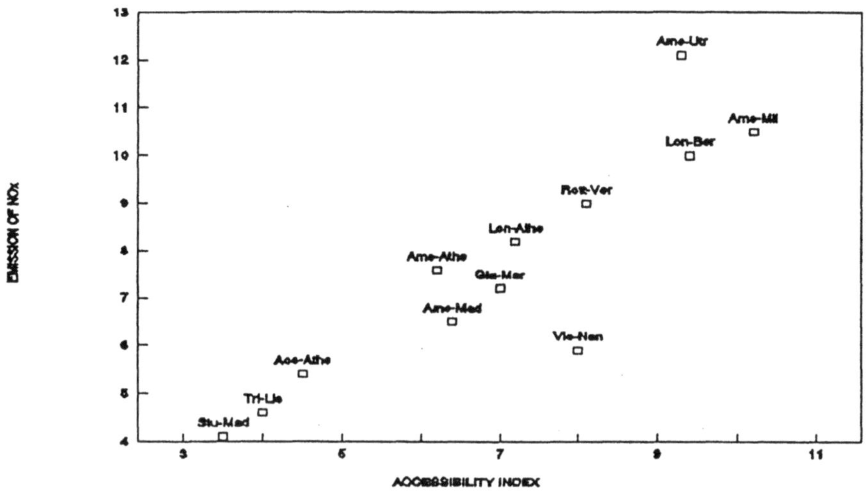


**Fig. 10.6.** Accessibility index (1988) versus total emission of NOx (1985)

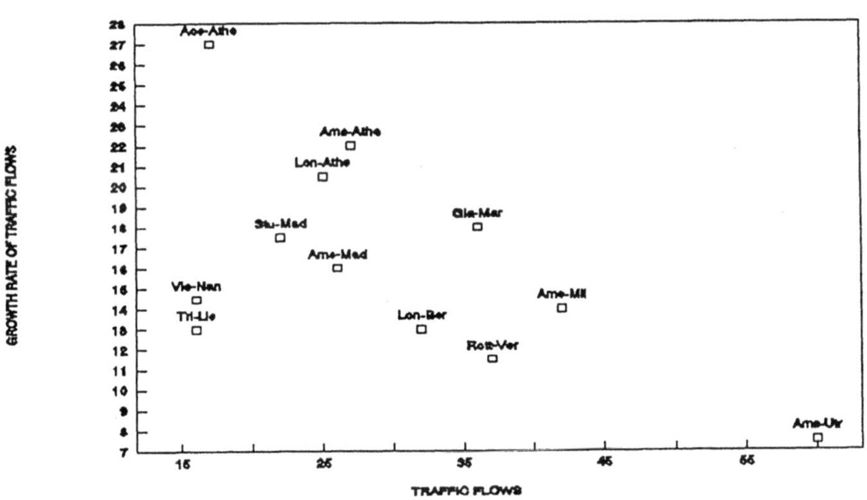


**Fig. 10.7.** Traffic flows (1985) versus traffic growth rate (1980-1985)

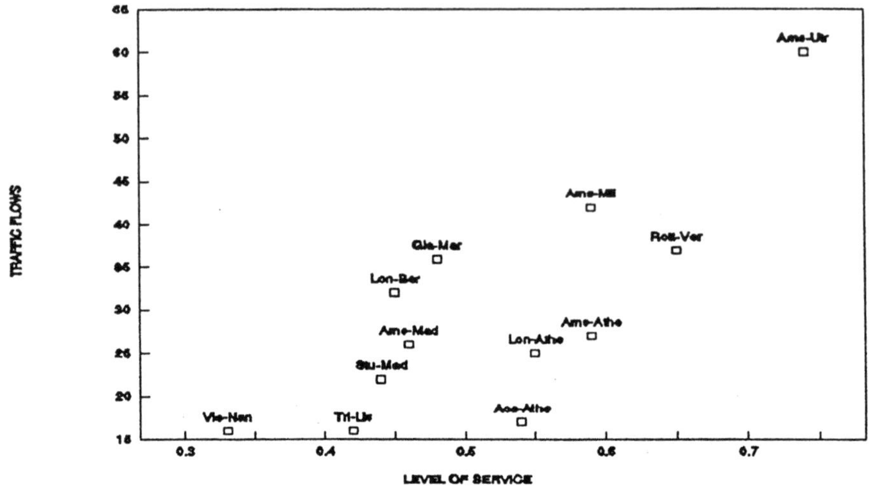


Fig. 10.8. Level of service (1985) versus traffic flows (1985) (in 1000 veh./day) on corridors

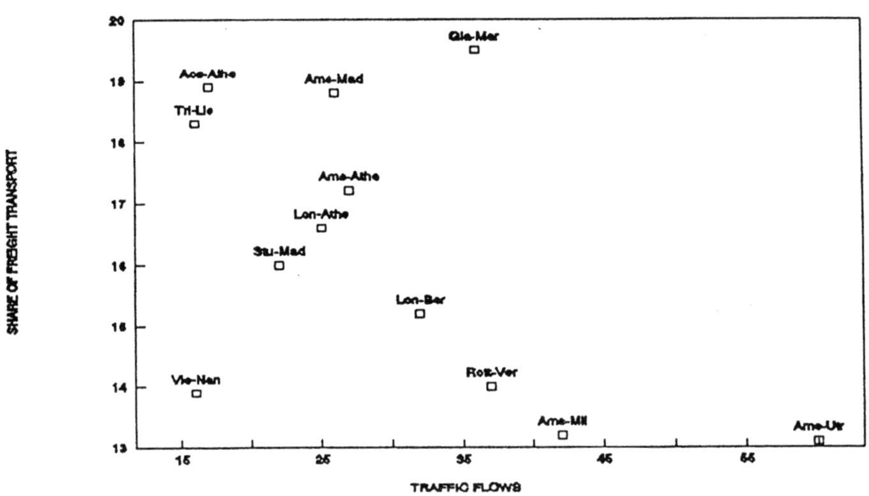


Fig. 10.9. Share of freight transport on total flows (1985)

All this information can be visualized in comprehensive diagrams (see Figures 10.10, 10.11, 10.12 and 10.13), where the role of the Aosta-Athens axis can be clearly identified in comparison with the other corridors.

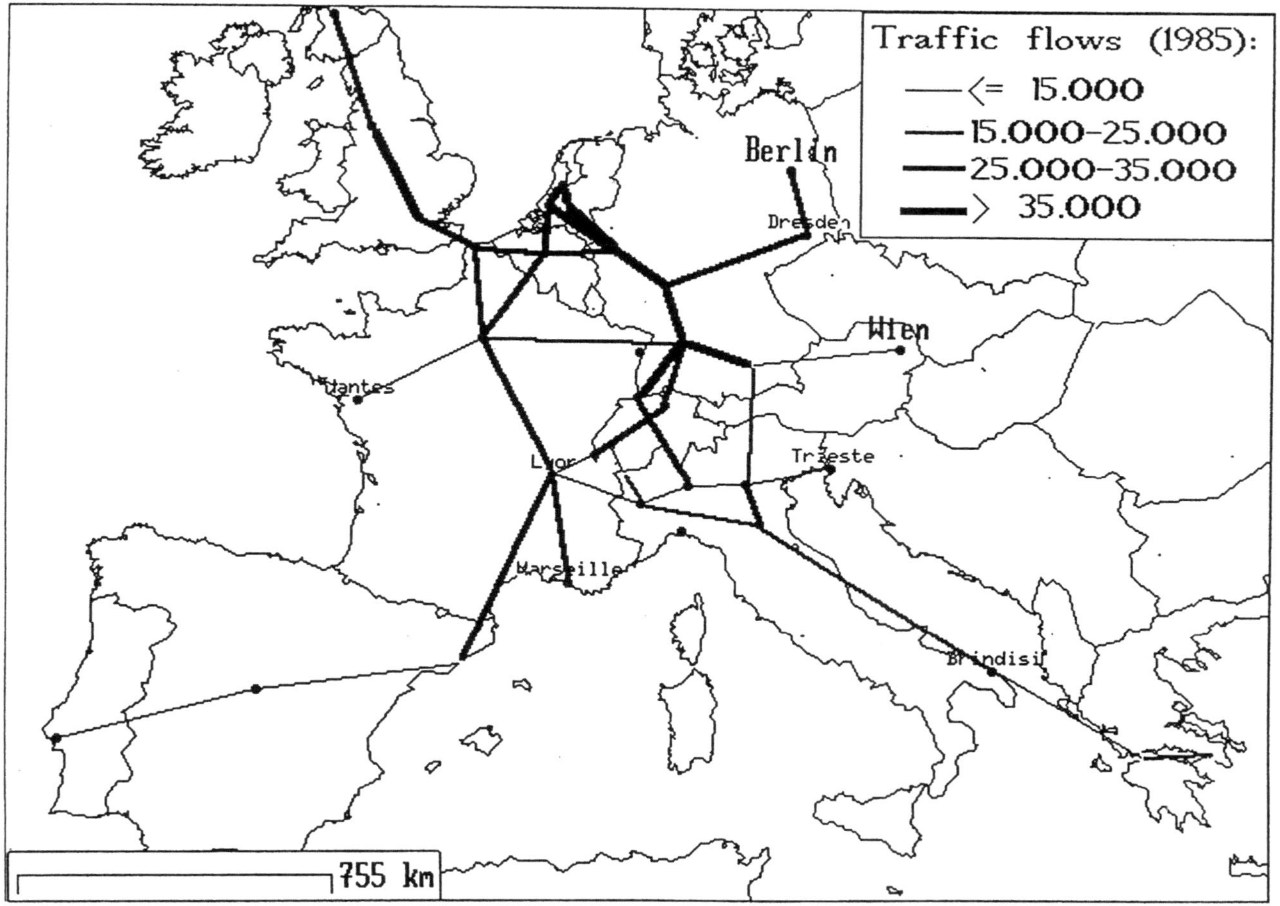


Fig. 10.10. Total traffic flows

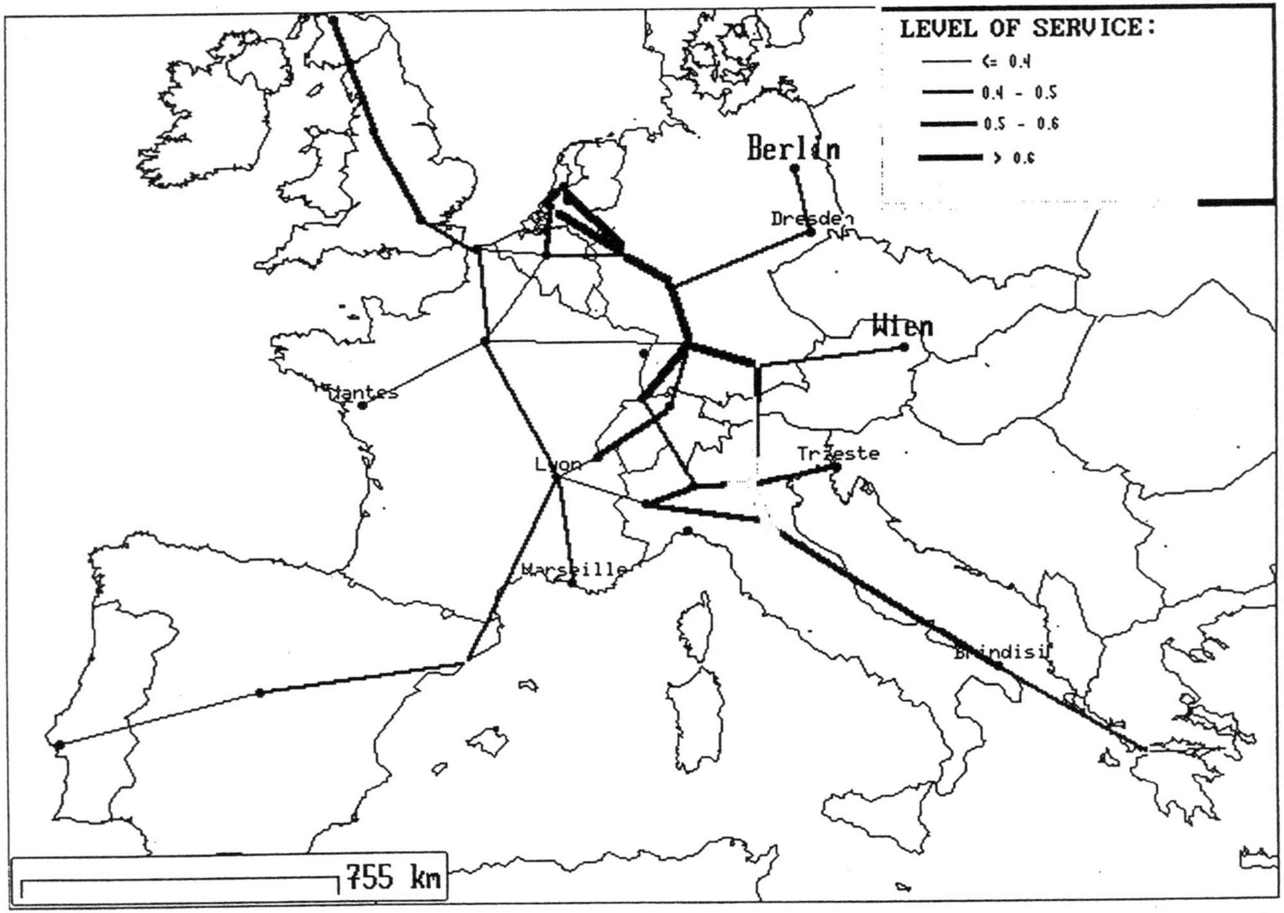


Fig. 10.11. Level of service along the segments

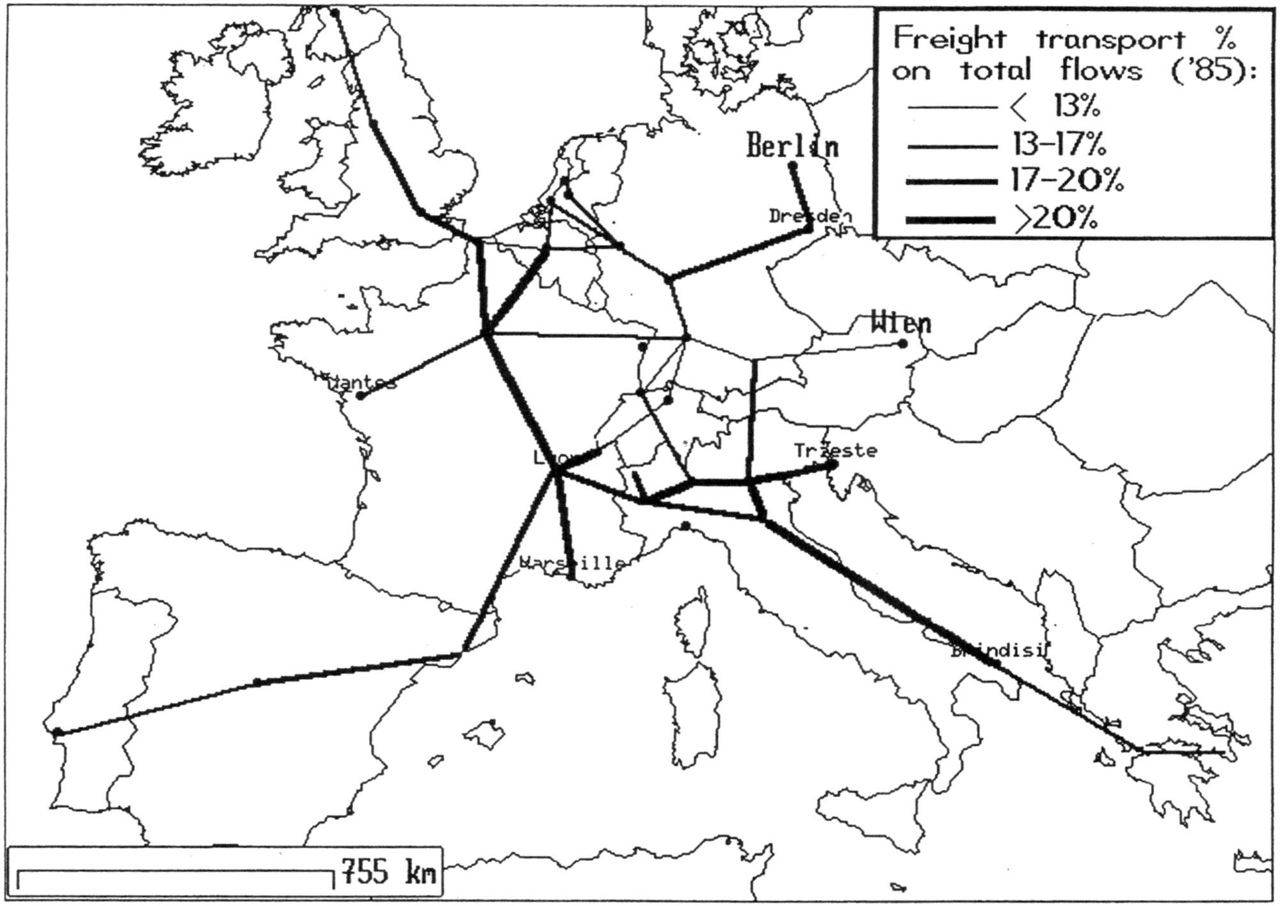


Fig. 10.12. Share of freight transport on total flows

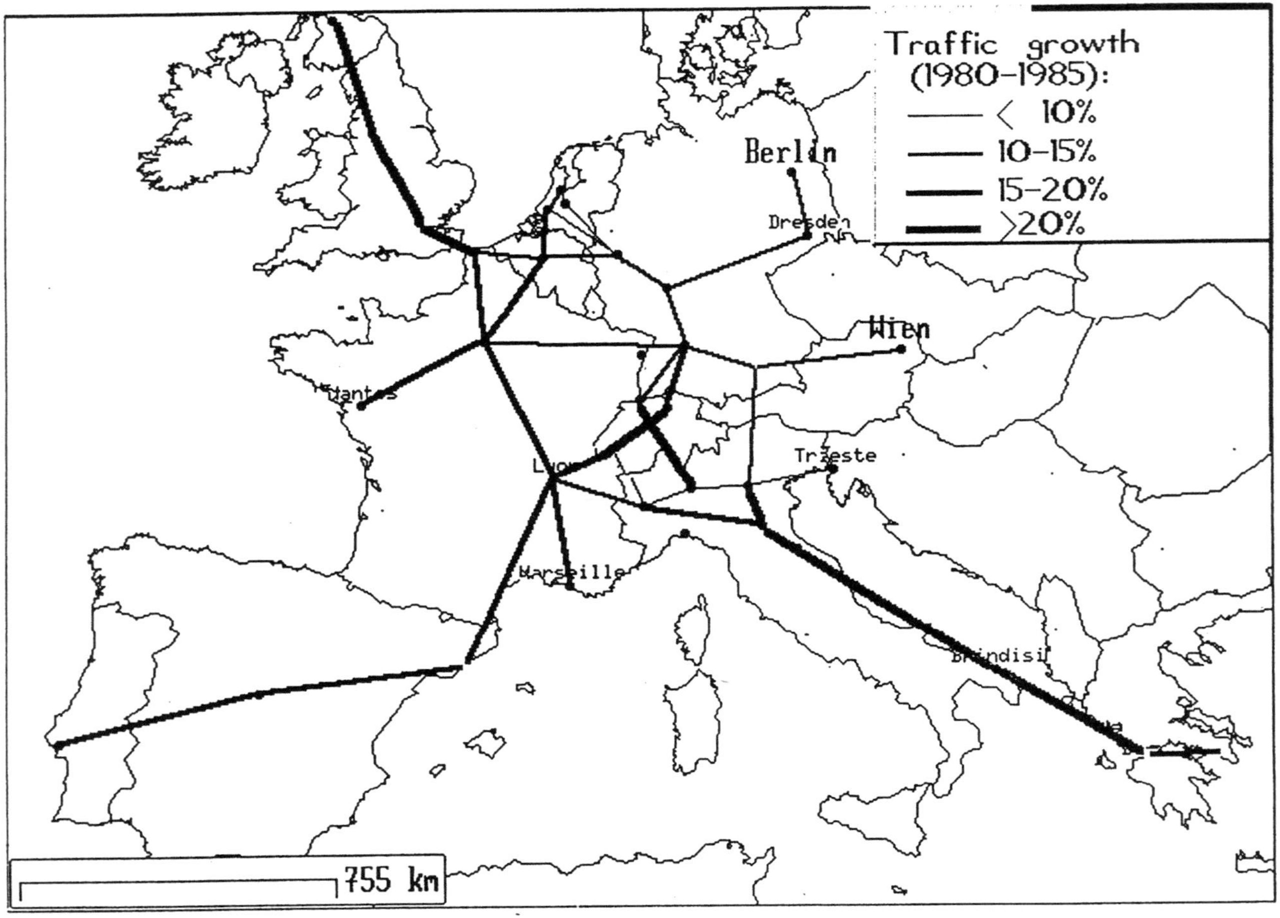


Fig. 10.13. Traffic growth rate (1980-1985)

In conclusion, on the basis of the above diagrams reflecting strategic information on socio-economic, transport supply and transport demand attributes, two main categories of European corridors emerge:

a) Central corridors:
Amsterdam-Utrecht
Amsterdam-Milan
Rotterdam-Verona
London-Berlin
London-Athens
Amsterdam-Athens

They are characterized by a high infrastructural density and accessibility degree, but are clearly insufficient in satisfying the total traffic flows (as witnessed by the high congestion level and pollution emission);

b) (Quasi)/Peripheral corridors:
Aosta-Athens
Glasgow-Marseille
Amsterdam-Madrid
Stuttgart-Madrid
Trieste-Lisbon
Vienna-Nantes

The axes are characterized by a medium congestion level and a low accessibility, and hence show the necessity of more infrastructural investment in order to avoid a rapid increase of congestion.

It should be noted that the above two-fold classification can be verified by means of a further analysis on interdependencies among the attributes, based on a Multi-criteria approach. This analysis will be presented in the next section. However, from our first examination it seems plausible that for each of the two groups of corridors, significant differences in both the potential of ATT introduction (as well as of other European resolutions) and the related behavioural responses are likely to emerge.

## 10.5 A Two-Fold Classification of Inter-Urban Corridors by Means of Multicriteria Analysis

### 10.5.1 The Methodology Adopted

In the previous section we have shown how the investigation of the corridors on the basis of the data collected has led to a preliminary typology, clustering the corridors into two main 'structural' groups (central and peripheral corridors). The above typology derives from the analysis of two-dimensional diagrams connecting the following types of variables for the corridors:

a) Socio-economic attributes
b) Transport supply attributes
c) Transport demand attributes

However, it should be noted that all these indicators constitute a *multi-dimensional profile* of the corridors which provides different aspects of the role of the corridors in relation to the potential introduction of ATT. This implies that the corridors can also be valued by a multi-attribute utility approach, by considering the previous attributes a), b), c) as a basis for a comprehensive evaluation of the corridors.

In this context, a multidimensional approach such as Multicriteria Analysis (MCA) is suitable for designing a ranking scheme of corridors in relation to their 'compound value', determined by the interdependencies of the various attributes.

**Table 10.5.** Criteria used in the multicriteria analysis

| | |
|---|---|
| a. Socio-economic profile | c. Transport-demand profile |
| a1 Population density | c1 Private car ownership |
| a2 (Total)GVAMP (*) | c2 Road traffic accidents |
| a3 GVAMP of industrial sectors | c3 NOx pollutions |
| a4 GVAMP of market services | c4 Vehicles flows |
| a5 Employment | c5 Growth of vehicle flows |
| a6 Industrial branches employment | c6 Heavy vehicle flows |
| a7 Market services br. employment | c7 Traffic flows smoothness |
| a8 Stock of dwellings | |
| b. Transport-supply profile | |
| b1 Motorway density | |
| b2 Accessibility of cities | |
| b3 Large agglomeration on corridor | |
| b4 Crossed EC member states | |

(*) GVAMP = Gross value added at market prices

MCA has become a popular tool in policy evaluation studies in many countries (see, e.g., Nijkamp, 1979 and Nijkamp et al, 1990), not only in the case of 'hard' (quantitative) information but also in the case of 'soft' (qualitative) information. In this framework, a suitable multiple criteria method, viz. the 'Regime Method' meeting methodological soundness, mathematical and statistical accessibility and easy computer use (Hinloopen et al, 1983; Hinloopen and Nijkamp, 1990) has been used in the present study. In our investigation, we assume *three main evaluation criteria*, viz. the Socio-Economic Profile, the Transport Supply Profile and the Transport Demand Profile. Each of these main judgment profiles is composed of observable criteria or attributes, based on the data collected, as shown in Table 10.5. Nineteen indicators, grouped in the three main criteria categories, are used here. The resulting impact matrix is shown in Table 10.6.

The application of the Regime Method is based here on five different scenarios, A, B, C, D and E, approximated by five different weight sets. Table 10.7 shows for each of these five scenarios an ordinal ranking of attributes according to their expected relevance, utilized in the MCA.

**Table 10.6.** The impact matrix

| | Socio-Economic Profile | | | | | | | | Transport Supply Profile | | | |
|---|---|---|---|---|---|---|---|---|---|---|---|---|
| | a1 | a2 | a3 | a4 | a5 | a6 | a7 | a8 | b1 | b2 | b3 | b4 |
| Amstedam-Utrecht | 12 | 8 | 6 | 11 | 12 | 10 | 12 | 12 | 12 | 10 | 1 | 1 |
| Amsterdam-Milan | 11 | 12 | 11 | 12 | 10 | 11 | 10 | 11 | 11 | 12 | 6 | 10 |
| London-Berlin | 10 | 9 | 8 | 8 | 11 | 12 | 11 | 10 | 10 | 11 | 2 | 9 |
| Rotterdam-Verona | 9 | 10 | 12 | 9 | 9 | 9 | 9 | 9 | 9 | 9 | 5 | 6 |
| Amsterdam-Athens | 7 | 6 | 9 | 5 | 7 | 7 | 6 | 7 | 7 | 4 | 11 | 11 |
| London-Athens | 8 | 5 | 7 | 6 | 8 | 8 | 8 | 8 | 8 | 7 | 12 | 12 |
| Glasgow-Marseilles | 6 | 7 | 5 | 7 | 6 | 5 | 7 | 6 | 4 | 6 | 8 | 2 |
| Amsterdam-Madrid | 3 | 4 | 3 | 4 | 3 | 4 | 4 | 3 | 3 | 5 | 9 | 5 |
| Aosta-Athens | 5 | 1 | 1 | 1 | 4 | 3 | 3 | 4 | 5 | 3 | 3 | 3 |
| Trieste-Lisbon | 2 | 2 | 2 | 3 | 2 | 2 | 2 | 2 | 2 | 2 | 10 | 7 |
| Stuttgart-Mardrid | 1 | 3 | 4 | 2 | 1 | 1 | 1 | 1 | 1 | 1 | 4 | 8 |
| Vienna-Nantes | 4 | 11 | 10 | 10 | 5 | 6 | 5 | 5 | 6 | 8 | 7 | 4 |

| | Transport Demand Profile | | | | | | |
|---|---|---|---|---|---|---|---|
| | c1 | c2 | c3 | c4 | c5 | c6 | c7 |
| Amsterdam-Utrecht | 12 | 12 | 12 | 12 | 1 | 2 | 12 |
| Amsterdam-Milan | 11 | 10 | 10 | 11 | 3 | 1 | 10 |
| London-Berlin | 10 | 11 | 11 | 8 | 4 | 5 | 4 |
| Rotterdam-Verona | 9 | 4 | 9 | 10 | 2 | 4 | 11 |
| Amsterdam-Athens | 7 | 6 | 7 | 6 | 11 | 8 | 9 |
| London-Athens | 8 | 9 | 8 | 5 | 10 | 7 | 8 |
| Glasgow-Marseille | 6 | 7 | 6 | 9 | 9 | 11 | 6 |
| Amsterdam-Madrid | 3 | 5 | 4 | 7 | 7 | 10 | 5 |
| Aosta-Athens | 4 | 8 | 3 | 3 | 12 | 12 | 7 |
| Trieste-Lisbon | 2 | 3 | 2 | 2 | 5 | 9 | 2 |
| Stuttgart-Madrid | 1 | 1 | 1 | 4 | 8 | 6 | 3 |
| Vienna-Nantes | 5 | 2 | 5 | 1 | 6 | 3 | 1 |

**Table 10.7.** Ordinal ranking of the attributes for five scenario sets A, B, C, D and E

| | |
|---|---|
| Scenario A | c7 c4 c5 c6 b3 b4 c3 b1 c2 b2 c1 a1 a8 a2 a5 a3 a4 a6 a7 |
| Scenario B | a7 a6 a4 a3 a5 a2 a8 a1 c1 b2 c2 b1 c3 b4 b3 c6 c5 c4 c7 |
| Scenario C | c4 c3 c7 c5 c6 c2 c1 b2 b4 b3 b1 |
| Scenario D | b1 b3 b4 b2 c1 c2 c6 c5 c7 c3 c4 |
| Scenario E | a1 a8 a2 a5 a3 a4 a6 a7 |

In the scenarios A and B, priority is given to attributes related to the 'traffic' criteria (both supply and demand) and to 'socio-economic' criteria, respectively. In the scenarios C and D the 'socio-economic' criteria are not taken into account, so that the corridors can be classified only in the context of transport characteristics. In scenario C more weight is given to the 'demand-side' attributes, in scenario D to 'supply-side' attributes. Finally, in scenario E priority is only given to socio-economic data, in order to evaluate the relationship between 'transport data' and 'socio-economic' characteristics. For each scenario a Regime analysis is carried out. The results are discussed in the next subsection.

### 10.5.2 Results and Conclusions

The results of the MCA (by applying a Regime Method) within the framework of the 'complex' evaluation of twelve European inter-urban corridors for the ATT introduction are presented in Table 10.8.

**Table 10.8.** The results of multicriteria analysis

| A | B | C | D | E |
|---|---|---|---|---|
| Amsterdam-Utrecht | Amsterdam-Utrecht | Amsterdam-Utrecht | Amsterdam-Milan | Amsterdam-Utrecht |
| Amsterdam-Milan | Amsterdam-Milan | Amsterdam-Milan | London-Berlin | Amsterdam-Milan |
| London-Berlin | London-Berlin | London-Berlin | Amsterdam-Utrecht | Lodnon-Berlin |
| Rotterdam-Verona | Rotterdam-Verona | Rotterdam-Verona | London-Athens | Rotterdam-Verona |
| London-Athens | London-Athens | London-Athens | Amsterdam-Athens | London-Athens |
| Amsterdam-Athens | Amsterdam-Athens | Amsterdam-Athens | Rotterdam-Verona | Amsterdam-Athens |
| Glasgow-Marseille | Glasgow-Marseille | Glasgow-Marseille | Glasgow-Marseille | Glasgow-Marseille |
| Aosta-Athens | Vienna-Nantes | Aosta-Athens | Aosta-Athens | Vienna-Nantes |
| Amsterdam-Madrid | Amsterdam-Madrid | Amsterdam-Madrid | Amsterdam-Madrid | Aosta-Athens |
| Vienna-Nantes | Aosta-Athens | Trieste-Lisbon | Vienna-Nantes | Amsterdam-Madrid |
| Trieste-Lisbon | Trieste-Lisbon | Stuttgart-Madrid | Trieste-Lisbon | Trieste-Lisbon |
| Stuttgart-Madrid | Stuttgart-Madrid | Vienna-Nantes | Stuttgart-Madrid | Stuttgart-Madrid |

From these results we can draw the conclusion that the *two-fold classification constructed in Section 10.4 seems fairly robust when confronted with changes in the scenarios: this is also the case for the MCA results.* In other words, our Multicriteria analysis displays the highest scores for the central group of corridors in light of relevant 'socio-economic and traffic' aspects. Thus, again a clear subdivision between central and (quasi-)peripheral corridors appears to emerge. It should be emphasized here that the rationale behind the definition of such a classification is the hypothesis that *differences in the impact of the corridors with ATT introduction are dependent upon hierarchical levels based on 'socio-economic and traffic' characteristics.*

## Acknowledgement

Research funds from the Italian C.N.R PFT2 - coutzect no 94.01441.PF74 are gratefully acknowledged. Although the work is attributable to all the authors, A. Reggiani has written Section 10.5. The contribution by G. Lampugnani is gratefully acknowledged. The remaining sections have been jointly written by all the authors.


## References


Hinloopen E, Nijkamp P (1990) Qualitative Multiple Criteria Choice Analysis. The Dominant Regime Method. Quality and Quantity 24:37-56

Hinloopen E, Nijkamp P, Rietveld P (1983) Qualitative Discrete Multiple Criteria Choice Models in Regional Planning. Regional Science and Urban Economics 13:77-102

Keeble D, Owens PL, Thompson C (1982) Regional Accessibility and Economic Potential in the European Community. Regional Studies 16:419-432

Nakicenovic N (1992) The Role of Surface Transportation. History of Development and Prospects for the Next Century. Paper presented at the Course on 'Communication Networks'. Summer University of Southern Stockholm, Novum Park

Nijkamp P (1979) Multidimensional Spatial Data and Decision Analysis. Whiley, New York.

Nijkamp P, Rietveld P, Voogd H (1990) Multiple Criteria Analysis in Physical Planning, North Holland. Amsterdam

Nijkamp P, Pepping G, Reggiani A, Lampugnani G (1993) European Corridor Typology. Description of Methodology and Data Base. Marta Working Paper No 13. EC-DRIVE, Brussels

Reggiani A, Lampugnani G, Nijkamp P, Pepping G (1993) European Corridor Characteristics. Some Typologies by means of European Maps and Diagrams. Marta Working Paper No 12. EC-DRIVE, Brussels

Reggiani A, Lampugnani G, Nijkamp P, Pepping G (1994) Towards a Typology of European Interurban Transport Corridors for Advanced Transport Telematics Applications. Paper presented at the Western Regional Science Association Meeting. Tucson, Arizona

# 11 Bottlenecks In Trans Alpine Freight Transport: A Multicriteria Analysis on Future Brenner Corridor Alternatives

P. Freudensprung, P. Nijkamp and R.-J. Simons

## 11.1 Introduction

Isolation refers to the limited accessibility of certain areas. Physical or other bottlenecks may act as barriers in the communication and transport systems connecting such areas. The Alpine region is a clear example of this phenomenon. Because of its natural conditions, the Alpine region offers only limited options and limited capacity to the transport sector. The geographic location of the mountain range in the middle of Europe separating important economic regions leads to the necessity for commodity flows through only a few corridors in this area. With the rapid increase of freight traffic during the last two decades, it has been difficult to find a satisfactory solution for both overcoming this bottleneck in the European transport network and guaranteeing the Alpine population an acceptable quality of life. Therefore, the problems of this part of the European transport network seem to be one of the most challenging ones to be solved.

Two major transit corridors for freight transport in a north-south direction can be identified on the Austrian territory:

- The 'Brenner-route', a corridor from Kufstein to Brenner, passing through the Inntal.
- The so-called 'Balkan-route' via Salzburg and then splitting into two corridors, the Phyrn and the Schober.

This paper will concentrate on the Brenner which is by far the most important transport corridor in Austria. In 1987, 62 million tons of freight were shipped through the Alpine region from the north of Europe to Italy. 20.75 million tons (16.45 million of tons truck traffic) crossed the Alps via the Brenner. These simple figures illustrate that congestion and environmental pollution are severe problems due to the high concentration of transport activity on this route.

The prognoses for the transport volume on the Brenner corridor vary, but all expect an annual growth rate of between 2% and 3.5%, that is a commodity flow ranging from 53 to 70 million tons per year in 2015. Consequently, an acceptable compromise between the goal of the Austrian (Tyrolean) government to strictly reduce the negative effects of transit traffic and the benefits for the EU-regions from an extended infrastructure capacity has to be found.

Although the region of the Alps offers only little available space to use, the maximum physical capacity seems not yet to be reached. The implementation of additional transport infrastructure (tracks, roads) is still feasible. But more important than the physical capacity is the ecological capacity, which has a far lower threshold level.

The environmental damage and the rising opposition of citizens lead to the conclusion that the ecological capacity limit has been exceeded under the euphoria of growth during the past two decades (Meadows 1972 1992). If this is the case, there are only two basic possibilities to solve the problem in a satisfactory way: either to reduce transport to a level below the ecological limit, or - since the physical capacity limit is also a logistic function - to push this limit towards a higher level by means of environmental friendly technological change. The best effects will be gained if the two ways are jointly used.

In the next section we will discuss in detail the so-called Pentagon model, which is a systematic approach to identify the critical success factors for an efficient and effective policy implementation and operation of transport systems. This approach will be used as the main frame of reference in this paper leading into the use of a multi-criteria analysis for evaluating different transport system alternatives for the Brenner corridor.

## 11.2 The Pentagon Model of Critical Success Factors

The Pentagon of critical success factors (Nijkamp and Vleugel 1993) was introduced to identify the necessary conditions for advanced new infrastructure networks (see Figure 11.1). In the sequel the five dimensions of the model will be used as a basis to identify the specific impediments to the Alpine freight transport on the Brenner route and to offer systematically possible interesting solutions. We will first outline the five essential elements of the Pentagon approach.

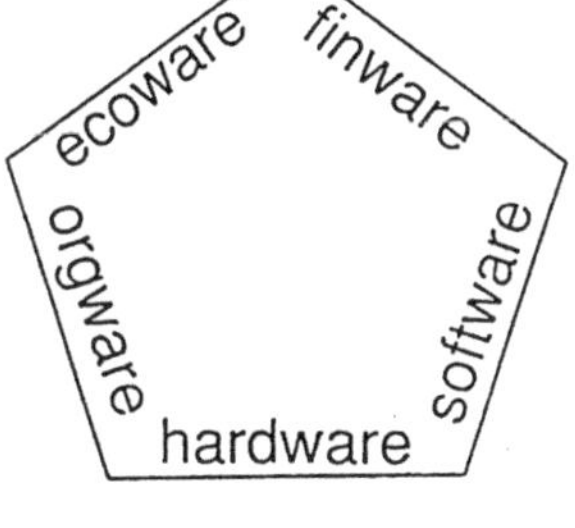


**Fig. 11.1.** The Pentagon model

### 11.2.1 Hardware

Hardware refers to the physical technological aspects of transport, for both road transport and railways.

*Road transport.* The congestion caused by the limited capacity of the four lane Brenner motorway can of course be alleviated by adding an additional lane. This may be realized by either constructing a new lane in both directions or by using the existing emergency lanes.

Furthermore, the use of trucks with a higher load may be seen as a way to reduce the number of vehicles. However, in contrast to current practice in many countries where trucks have a maximum gross vehicle weight of 50 tons, the weight limit for Austria is 38 tons. On the other hand, large vehicles cause greater damage to the motorway and hence cause higher maintenance costs.

*Railways.* The renewal of the Austrian railway tracks (with a designed speed of up to 160 km/h) will be completed in 1997. Additionally, a tunnel for freight trains to by-pass Innsbruck is due to be opened in order to reduce the travel time through the corridor of the Unter Inntal. The capacity added by this project will be 70 trains per day. The planned Brennerbasistunnel and the adjustment of the tracks for a high speed railway system (with a designed speed of 200 km/h) will double the daily capacity of transit trains. Not only the tracks, but also the rolling stock have to be adjusted in order to meet the needs of modern transport systems. The necessity to cooperate with the road sector in terms of intermodality and the increasing demand for individual containers of relatively small shipments (the trend of dematerialization) needs the use of both flexible and standardized wagons.

To favour a shift from road to rail on the Brenner corridor, first the 'Rollende Landstraße' was introduced, where the truck tractor plus semitrailor are loaded on the train (cf. Backhausen 1992). Due to the uneconomical high percentage of dead weight, this operation is not likely to survive in the long run. Containers, swap bodies and the road railer have become more promising projects to invest in.

### 11.2.2 Software

Software refers to information and use of measures that serve to increase the efficiency of transport operations. Traffic management measures taken in the past dealt mainly with time and access restrictions for the Brenner corridor. Though there is also the possibility to restrict passenger cars (e.g., to guarantee a constant speed for trucks), only trucks have been affected by restrictions so far, mainly because of ecological reasons. All Austrian roads are subject to a night-time ban (22.00-6.00 hrs.) and a ban on Sundays and holidays for trucks.

American studies have demonstrated that the cost effectiveness of road investment increases if an additional lane is exclusively reserved for trucks. There is no operational experience with truck-only lanes yet, but this might be a step towards improving the performance of the Brenner motorway. The main obstacle for rail transport is caused by the relatively long operating times of transshipment. Therefore, investment in intermodal terminal infrastructure is necessary to guarantee an intermodal competitive edge for freight transport by rail. The introduction of advanced information systems will be another decisive method to improve both rail and road transport, with a major focus on combined transport. Such information systems may concern:

- road conditions and traffic situations
- free capacities on both rail and road
- type and volume of commodities to be shipped.

It is important to have information systems for both the users and the operators in a transportation network. While the first alternative may also be included in traffic guidance systems (e.g. LISB, Prometheus) and is of main interest to forwarders to avoid congestion on the road, the latter is more concerned with the economic efficiency of shipping. Better information about where and when to find freight (capacity) will accelerate transport operations and increase the efficiency of the network.

### 11.2.3 Finware

Finware is concerned with economic and financing aspects of transport. In the 1980s the total investment in transport infrastructure was being decreased, although the volume of traffic was rapidly growing. So the capacity constraints arising nowadays are a logical consequence of this past policy. In order to finance expansion of infrastructure networks, user charges are often advocated. This means that all relevant costs of operating and managing infrastructure have to be directly borne by the user of the transport facilities. This is important, as transport prices do not include all costs caused by the various transport modes. For example, road infrastructure is available to freight transport without proper compensation for the higher construction costs of motorways that have to be suitable for vehicles of larger dimensions and weight than passenger cars. This holds also for the high maintenance costs, which are to some 90% caused by trucks. This biased price system thus offers road transport an important advantage in competition with respect to rail transport.

The motorway from Innsbruck to Brenner is at present a toll road; and most likely this toll road system will be extended in the near future.

This brings us to the issue of financing the infrastructure. Infrastructure investments bear a relatively high risk-benefit ratio. The combination of high political risk, a long pay-back period (mostly >20 years) and relatively low revenues appear to discourage private investors. Traditionally, investments in infrastructure belonged to the public domain, because the goal was also to guarantee a public service to peripheral regions by using the revenues from profitable parts of the networks. In any case, private financing is gaining importance (see for instance Nijkamp and Rienstra 1995). Thus, there may be a fair chance to finance the new infrastructure of the Brenner corridor from private sources, in particular in light of the growing public deficit in Austria.

A more practicable method which is gaining importance in many European railway companies might be to divide the infrastructure and the operational sector of the railways, as is also done for roads. The right to use a specific corridor at a specific time may then be sold to operators by way of a concession. This slot-exchange is already used in air traffic and may also be practicable for trains. Moreover, the price mechanism implied in this system might then be used to regulate the congestion in a high traffic area like the Brenner.

### 11.2.4 Ecoware

Ecoware refers to the environmental conditions to be fulfilled by transport. In Austria nowadays, the environmental impact of traffic in general and transit traffic in particular is a hot item. In fact, there are two kinds of problems in relation to the Alps. First, there is the 'natural-environmental' problem because of the environmental vulnerability of the Alps. This means, for instance, that the mountain forests are relatively quickly affected, while it takes a long time to recover. The mountain-forests are very important to protect the mountain against erosion, but also to protect the people against avalanches.

Secondly, there is, because of the Alps, very little space to build more roads or railways. In fact, there are only a few places where one can pass the Alps by car. This means that the traffic passing the Alps is heavily condensed on a few roads (for example, the Brenner-pass). An alternative may be to build a tunnel, but this is often expensive.

### 11.2.5 Orgware

Orgware deals with the control and organization of infrastructure. In trying to control and influence the actual and future traffic flows, the Austrian government is changing the organization of infrastructure. The old situation is comparable to that of many European countries. The use of infrastructure is indirectly subsidized, because the drivers do not pay the full price for the infrastructure used. Secondly, about 20% of the trucks is empty because of cabotage agreements and the network is not functioning optimally because of a separate development of infrastructure.

In order to solve these problems the Austrian government recognizes that it has to follow the way of most European countries. First, in the coming years new measures are to be expected that will make the railway system more competitive, for instance, more market orientation.

Secondly, the Austrian government will gradually introduce a toll system to let transit trucks pay for the infrastructure used.

Thirdly, there is a great interest in reducing the amount of empty trucks, because that will also reduce the total amount of traffic.

Having now briefly discussed the five components of the Pentagon model, we will in the next section assess and evaluate different Brenner corridor options.

## 11.3 Evaluation of Brenner Corridor Alternatives by Means of Multicriteria Analysis

A multicriteria analysis aims to identify the most plausible realistic option out of a set of discrete alternatives, by investigating in a systematic manner all relevant impacts of each alternative on the basis of a set of evaluation criteria. This means that the construction of a systematic impact table is an important step (see Nijkamp and Blaas 1994).

### 11.3.1 Alternatives

At present, the infrastructure of the Brenner corridor consists of a motorway with four lanes (two in each direction) from Kiefersfeld to the Brenner, with 3,700 trucks (1991) passing each day. The construction of a third lane has partly been realized, as it was mainly built as a 'creep lane' to guarantee smooth traffic on steep parts. The completion of the third lane over the full length of the motorway may be realized in the medium-term future, although resistance from the local population may be expected.

The alternatives given for improving the railway network on this route are backed by both the population and the government. Budget restrictions will have a major influence on the level of improvement that will be realized and in which time span. This means that various alternatives may be envisaged. The following alternatives are possible depending on a number of factors, in particular the political pressure on the Austrian government from both the environmentalists and the industry. Their lobbying within the EU will be of decisive importance.

| | |
|---|---|
| Alternative I | Renewal and doubling of the tracks in the Unter Inntal and construction of a third lane on the motorway. |
| Alternative II | Renewal and doubling of the tracks in the Unter Inntal but no capacity increase of the road infrastructure. |
| Alternative III | Brenner-Basistunnel and construction of a third lane on the motorway. |
| Alternative IV | Brenner-Basistunnel but no capacity increase of the road infrastructure. |

These four alternatives will be evaluated by means of relevant assessment criteria which will now be successively discussed.

### 11.3.2 Criteria

The basic idea here is to operationalize different criteria according to the dimensions of the Pentagon model in combination with other important factors. Next, the design of scenarios (by changing the political weight factors of the assessment criteria) may then highlight the potential of the various alternatives under varying preconditions. Due to lack of data, here we will mainly use a qualitative multicriteria analysis. The criteria taken into consideration will first successively be discussed (see for more details Simons and Freudensprung 1994)

*Land use.* This is an important factor, as human activities require more and more space. Space is a scarce good in an Alpine valley and has to be used with caution. Therefore, land use as a result of new infrastructures is an important policy criterion.

*Energy consumption.* The factor used to measure the energy consumption of the two transport modes, namely road and rail, refers here to the transportation operation only. Of course, it would be better also to calculate the energy used during the construction of infrastructure and the rolling stock in order to have a complete eco-budget, but this is problematic because of data reasons.

*Pollution/emission.* Transport causes several emissions. Here we will only look at freight transport. The table of Kürer (ECMT 1992) is used here, as this contains the most important emission of traffic, namely CO2. As we want to use the different emissions in one criterion, the total sum of emitted g/tkm will be calculated over all the relevant substances.

*Project costs.* Of course the costs of investment for each alternative are a decisive factor in any evaluation. The budget situation of the Austrian government is subject to many constraints, as there are more and more fields competing for public funding. In our calculation only the construction costs are taken into account, as no data were available on the operational costs per ton for each of the four alternatives.

*Noise.* Noise is an often underestimated factor of environmental damage. Some 54% of the people feel directly affected by this type of pollution. The quantitative measurement of the disturbance caused by noise is very difficult, because it does not only depend on the volume (dB(A)), but on numerous other factors that can hardly be taken into account.

*Integration effects.* The economic effects of a higher accessibility for a region is a useful factor in order to judge the economic potential of infrastructure investments. In our case it was not possible to quantify the effects of the European integration for the area of interest because of lack of data, so that we had to resort to a qualitative assessment.

*Free railway capacity.* According to a study of the Austrian Traffic Ministry, the total volume of transit traffic in 2000 will be about 75% higher, compared to 1990. The motorway transport will grow by approximately 100%. The problem in our case study is to estimate the volume of traffic that will pass via the Brenner. This will of course depend on the various alternatives chosen. Our aim is to estimate the extent to which the measures of the Austrian government may create flexibility with respect to future transport developments. Thus we have to look at the free capacity, because in this way it is possible to influence the traffic flow in a more economic way. If, for instance, in the future the government wants to have more goods transported by trains, then it is necessary to have sufficient capacity left in the railway system. In this way we may define 'free capacity of the railway' as the free capacity that is not yet used.

*Safety.* Although a declining accident rate on the Brenner Corridor is noticed, there is still a big problem concerning traffic safety. Another problem is how to estimate safety. We can of course only count the fatal accidents, but that leaves out the 'normal' accidents which often have many consequences for the functioning of the transport network. To find an expression for fatality chances, we have to look into the differences between road and rail. Transit traffic on the road has on average a 60% higher risk of being involved in an accident. Furthermore, trucks on transit routes are relatively more involved in accidents than private cars. Finally, we know that road transport is about 10 times more risky than railway transport. Then we can use these global figures - in combination with traffic volumes - to assess the fatality risks in terms of ordinal ranking.

*Speed.* Of course, the criterion speed is very important. With speed we do not only mean the actual speed that can be achieved, but it is also an approximation of congestion. It is, for instance, plausible that the congestion on the Brenner will get worse, if the capacity of the network does not grow. On the other hand, according to the Austrian government, there will be less congestion in the railway system, because of the many great improvements to take place in the future. Thus, the average speed in any of the four options has to be assessed.

*Reliability.* Reliability refers to the robustness of estimated travel time with regard to external shocks, such as weather conditions. This can only be approximated in a qualitative sense.

*Landscape segmentation.* This is an environmental criterion. The area of the Brenner pass is a touristic, agricultural and yet densely populated area. It is therefore important to see in what way the landscape will be segmented by new infrastructure projects. In this case we will also use an ordinal approximation for our assessment.

Having discussed now eleven evaluation criteria for the four Brenner Corridor alternatives, we may construct a comprehensive impact matrix which comprises in a numerical form (quantitatively or qualitatively) the various outcomes. This information is contained in a concise form in Table 11.1.

**Table 11.1.** The comprehensive impact matrix for the Brenner Corridor alternatives.

| | Alternatives | | | | |
|---|---|---|---|---|---|
| criterion | I | II | III | IV | units of measurement |
| 1 | 12 | 9.1 | 12.02 | 9.12 | $km^2$ |
| 2 | 43.14 | 34.47 | 47.69 | 39.02 | n/MJ |
| 3 | 4130.6 | 3287.3 | 4506.8 | 3663.5 | g/tkm |
| 4 | 9 | 7 | 58 | 56 | billion AS |
| 5 | 1 | 3 | 2 | 4 | ordinal |
| 6 | 3 | 1 | 4 | 2 | ordinal |
| 7 | 16 | -157 | 216 | 45 | free train capacity |
| 8 | 3 | 4 | 2 | 1 | ordinal |
| 9 | 125 | 110 | 145 | 130 | average speed |
| 10 | 2 | 1 | 4 | 3 | ordinal |
| 11 | 1 | 2 | 3 | 4 | ordinal |

## 11.4 Scenarios for the Brenner Corridor

Now we turn to the specification of different policy scenarios which can be formulated by varying political weight or priorities attached to each of the successive criteria. Each scenario thus represents a specific political position or angle reflected in the fact that the criteria are weighted differently. We distinguish here the following six scenarios:

*Indifference;* all criteria are given the same weight.

*Ecology I;* all eco-factors are assumed to be more important than the other criteria.

*Ecology II;* here the eco-factors are split up into two groups; the first group consists of the components directly infringing on people, while the indirect eco-effects are assumed to be less important.

*Ecology-cost mix;* the same as Ecology II, but the construction costs are here more important than the indirect eco-criteria.
*Democracy;* the weight factors are ranked according to the hottest current political issues for the Austrian population.
*Network efficiency;* the most important criterion is the construction costs, followed by the criteria characterizing the network potential.
In Table 11.2 the ranking of the criteria for these six scenarios is presented.

**Table 11.2.** Brenner Corridor scenarios based on ranking of criteria.

| | Scenarios | | | | | |
|---|---|---|---|---|---|---|
| criterion | a | b | c | d | e | f |
| 1 | 1 | 2 | 2 | 2 | 2 | 1 |
| 2 | 1 | 2 | 2 | 2 | 2 | 1 |
| 3 | 1 | 2 | 3 | 4 | 3 | 1 |
| 4 | 1 | 1 | 1 | 3 | 3 | 3 |
| 5 | 1 | 2 | 3 | 4 | 3 | 1 |
| 6 | 1 | 1 | 1 | 1 | 1 | 2 |
| 7 | 1 | 1 | 1 | 1 | 1 | 2 |
| 8 | 1 | 1 | 2 | 2 | 3 | 1 |
| 9 | 1 | 1 | 1 | 1 | 1 | 2 |
| 10 | 1 | 1 | 1 | 1 | 1 | 2 |
| 11 | 1 | 1 | 3 | 4 | 2 | 1 |

The Tables 11.1 and 11.2 can now be further treated by applying a multicriteria analysis for discrete choice alternatives. Given the qualitative nature of these tables, the Regime method is particularly useful (see for all details Nijkamp et al 1991)
In Table 11.3 the results of the multi-criteria programme Regime are presented.
The results from Table 11.3 show that alternative II and IV are most favoured. This suggests that there should be no increase of road capacity, as suggested in the scenarios I and III. This is important to note because the Austrian government is planning more road capacity to solve the traffic problems on the Brenner in the medium term, while in the long run a Brenner-basis tunnel is planned. The plans of the Austrian government are in a way comparable to scenario III, which is, according to our analyses, the worst alternative.

**Table 11.3** Results of the Regime method

| | SCENARIO | | | | | |
|---|---|---|---|---|---|---|
| ranking | (a) | (b) | (c) | (d) | (e) | (f) |
| first | IV | II | IV | II | II | IV |
| second | II | IV | II | IV | IV | II |
| third | I | I | I | I | I | I |
| fourth | III | III | III | III | III | III |

Secondly, for both alternative II and IV the number of times they score the first place is equal and they change positions with each other. An explanation for this phenomena is found in Table 11.1. First, alternative IV seems to be some sort of compromise-alternative; there is no exceptional low or high score for any criterion. Alternative II is for a large part comparable to alternative IV where the low construction costs compensate for the strong network-performance. There seems to be a trade off between scenario II and IV. Scenario II is stronger in eco-factors and has a lesser score in network-performance, as scenario IV has a strong network-performance and a lesser score in eco factors. Alternative I and III seem to be more radical with bad scores on the eco-factors and only modest scores on the network-criteria.

## References

Backhausen F, Ewers HJ (1992) Marketingstrategien für den schienengebundenen Güterverkehr
BMöWV (1994) The Vienna Paper
BMöWV (1991) Das österreichische Gesamtverkehrskonzept
ECMT (1992) Transport growth in question. 12th Symposium Lisboa
Maggi, Nijkamp P (1992) Missing Networks in European Transport and Communication. European Planning Studies
Meadows (1992) Die Grenzen des Wachstums
Nijkamp P, Rietveld P, Voogd H (1991) Multicriteria Analysis in Physical Planning. Elsevier, Amsterdam
Nijkamp P, Vleugel P (1993) Success Factors for High Speed Rail Networks in Europe. International Journal of Transport Economics
Nijkamp P (1994) From Missing Networks to Interoperable Networks. The Need for European Cooperation in the Railway Sector. Transport Policy (forthcoming)
Nijkamp P, Rienstra S (1995) Private Sector Involvement in Financing Transport Infrastructure. Annals of Regional Science, 1 (forthcoming)
OECD (1992) Road transport research - cargo routes. Truck roads and networks
öIR (1989) Alpenquerender Transitverkehr.
Rietveld P, Nijkamp P (1992) Transport and Regional Development. Research Memorandum, Free University, Amsterdam

# 12 Trade Effects of the Emerging Market Economies: A Study of the Transport Potential of the Rhine-Main-Danube Waterway

M. Fischer and C. Rammer

## 12.1 Introduction

During the last few years, the economic landscape of Europe has changed dramatically, both in the East and in the West. In Eastern Europe, the peaceful revolutions - except Yugoslavia - are among the most dramatic events of the current epoch. A central element of the revolutions in most of these countries is a total reversal of economic policy and the transition from a centrally planned to a market-based economy. In Western Europe the process of economic integration within the EU has been and will be intensified with the implementation of the Economic and Monetary Union in 1993 and integrating the EFTA countries as most of them are likely to become EU-members in 1995. Certainly, both events will have far-reaching impacts on trade patterns and transport systems in Europe.

Today's liberalization and opening up of the emerging market economies in the East -including the adoption of hard currency accounting, investment by and joint-ventures with Western firms, industrial restructuring and privatization, the removal of trade barriers, the implementation of new technologies, and political deregulation - will affect East-West trade patterns in several aspects in Europe. First, the increase in openness and the expansion of output will positively affect the volume of East European countries' trade. Second, new demand structures, new production technologies, and new types of specialization will lead to remarkable changes in the composition of their trade patterns. Finally, the dismantling of CMEA trading arrangements and the integration of Eastern European countries into the European Economic Area will result in a spatial re-orientation of trade patterns (Collins and Rodrik 1991; Hamilton and Winters 1992; Rollo and Stern 1992).

There have been a number of recent studies on the implications for the world economy of developments in Eastern Europe (e.g. Collins and Rodrik 1991; Debs et al. 1991; Siebert 1991) and ours is naturally related to these. The major focus in this chapter is on the likely trade effects of the emerging market economies on trade within the Rhine-Main-Danube (RMD) area and the potentials for the RMD waterway transport. The RMD area represents the geographical framework of the study and is defined as the wider catchment area of the RMD waterway including

Austria, Belgium-Luxemburg, Bulgaria, the Czech and Slovak Republics, France, (West) Germany, Greece, Hungary, the Netherlands, Romania, Switzerland, Turkey and the former Yugoslavia.

The chapter is organized as follows. In Section 12.2 the process of economic transformation in Eastern Europe is briefly characterized by considering the major areas of reform and discussing the question of the future economic development of Eastern Europe. The focus of Section 12.3 is on the likely impacts of the opening up of Eastern Europe on East-West trade. The discussion is strongly based on a study exploring the potential volume, direction and composition of trade in the RMD area (Fischer and Rammer 1993; Fischer and Johansson 1994). Section 12.4 provides a quantitative assessment of the likely effects of the emerging trade patterns on goods transportation on the RMD waterway. Finally, some policy recommendations are made to tackle major bottlenecks of the RMD waterway.

## 12.2 The Economic Transition Process in Eastern Europe

Since 1989 Eastern Europe is experiencing a period of transformation from centrally planned economies towards open market economies which is unprecedent in history. There is a broad agreement on the major steps which have to be undertaken to transform a socialist economy into a market economy (Siebert 1991; Hinds 1991; Genberg 1992; Funke 1993). The major disagreements focus on the appropriate sequencing and speed with which these policy measures should be implemented. Figure 12.1 lists the key issues that have to be addressed in any transition. In this figure three types of measures are distinguished: institutional reforms, macroeconomic stabilisation and microeconomic reforms.

It is widely agreed that institutional reforms are central to creating a healthy market economy. Institutional changes focus on the contract and company law, property rights, two-tier banking system, the regulatory system etc. Extensive legal reforms are needed, including legislation to protect private property. A key issue is the development of a two-tier banking system with an autonomous and independent central bank and private commercial banks. The development of financial intermediaries which can channel domestic saving into needed investment projects is a further key issue.

There are a variety of microeconomic reforms that are essential to creating a healthy market economy. We mention here only a few reforms from a long list. First, enterprises have little incentive to respond to market signals when their outlays are simply financed by subsidies from the government. Thus, it is crucial to establish hard budget constraints for firms to weed out those that will be insolvent at realistic prices. Second, governments have to take steps to develop the private sector, including the privatization of state-owned enterprises and the elimination of restrictions on the establishment of private firms. The creation of new firms is an important aspect of restructuring the economy (free market entry of new firms). Third, a crucial issue is dismantling the complex system of centrally controlled prices and allowing them to reflect relative scarcities and tastes in the most important goods markets. One effective means of establishing a sensible set of relative prices, at least for tradable goods and services, is to import prices from the world economy through trade liberalization. Fourth, also a key issue is to attract foreign direct investment that brings new technology and knowhow as well as access to foreign markets.

| Institutional Reforms | | |
|---|---|---|
| Legal System | Reform of the Banking System | Regulatory System |

| Microeconomic Reforms | | |
|---|---|---|
| Development of the Private Sector | Price and Market Reforms | Trade Liberalisation |

| Macroeconomic Stabilisation | | |
|---|---|---|
| Currency Convertibility for Current Account Transactions | Fiscal and Tax Policies | Hard-Budget Constraints |

Source: Siebert 1991, Fischer and Gelb 1991, Dornbusch 1991

**Fig. 12.1.** Elements of transformation to a market economy. Three major areas of measures

Macroeconomic stabilization can not be too distant from the microeconomic reforms. Macroeconomic stability here means balancing total domestic demand with total domestic production in such a way that the price level is relatively constant and predictable, and that the country does not need to borrow from abroad beyond its future capacity to service its debt (Collins and Rodrik 1991). In the absence of such a stability, it is difficult for prices to act as effective market signals which is a crucial element of a market economy.

Achieving macroeconomic stability entails establishing and maintaining a fiscal policy -characteristically with the budget at or near balance - and control over the growth of the money supply. Budget control will require eliminations of many of the price subsidies and is, thus, tied to monetary adjustment over the medium run. Tax reforms will be essential because among other things government revenues are tied to a system of price controls in these economies. Opening up the economy can lead to large external imbalances if macroeconomic policies are unsustainable, or if the exchange rate is substantially overvaluated.

A central point in our study is to stress the heterogeneity of the Eastern European countries in the RMD area, i.e. Bulgaria, the Czech and Slovak Republics, Hungary, Romania and the former Yugoslavia. The countries differ in a number of important aspects. There are considerable differences in their sectoral composition, with agricultural production accounting for a large share of the total in countries such as Romania while others, such as the Czech and Slovak Republics, are considerably more industrialized. In the beginning of this decade, the RMD countries were at very different starting points in terms of how much relative prices and the allocation of resources differed from the outcomes which might be expected from functioning markets. Hungary, for example, had been liberalizing gradually for about two decades, whereas other countries such as Bulgaria or Romania had hardly started to initiate economic reforms. There are also differences in the prevailing political and social situations which can be a central force to push or hinder economic reforms. These conditions also contribute to the extreme uncertainty that is existing about the future course of events in some of the countries.

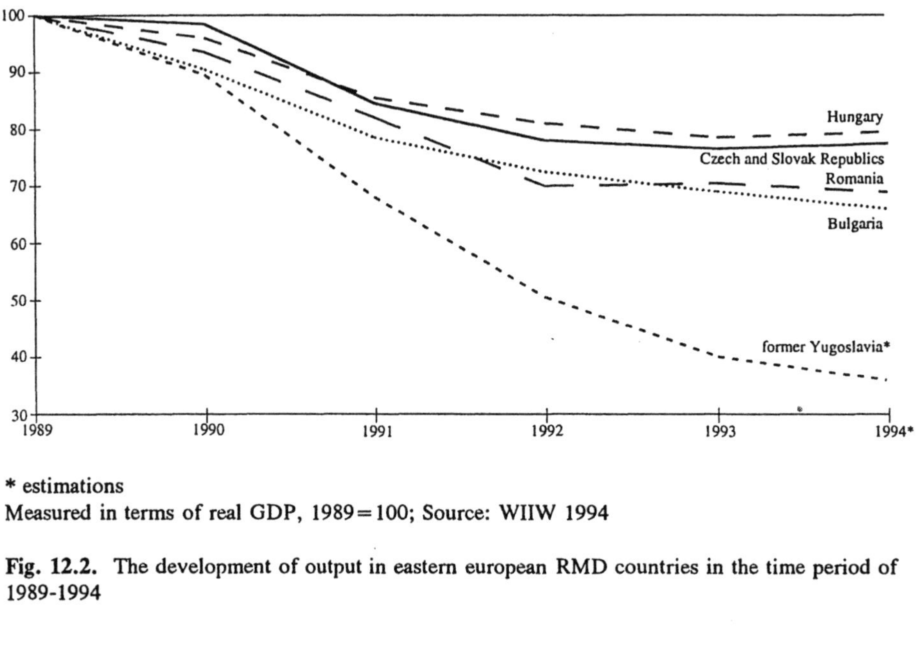


* estimations
Measured in terms of real GDP, 1989=100; Source: WIIW 1994

**Fig. 12.2.** The development of output in eastern european RMD countries in the time period of 1989-1994

Although each of the countries concerned has unique features, it is useful to identify a few broad problems. First, Bulgaria and Romania adopted radical reforms in 1991, but the outcomes remain in doubt. Second, Hungary and the Czech and Slovak Republics are in the kind of more gradual liberalization programs. Considerable progress has also been made in Hungary. The speed of change in the former Czechoslovakia (especially in the Czech Republic), accelerated in the early 1990s. Third, we know that the transition from a planned to a market economy is associated with a specific development in output - a J-curve in output with a dip and a valley, where output will fall first and then start to rise (Siebert 1991). The shape of this curve is not exactly known, but is relevant for the political economy of transition. The shape of the J-curve depends on the inefficiency of the existing firms, the speed and methods of privatization, the speed with which new firms come into existence, the success of enterprise restructuring etc. The statistical measurement of the J-curve is not without problems. For simplicity we use real GDP as measure of output in Figure 12.2. So far we can observe the falling branch of the J-curve in all cases. The data suggest that the fall of output might come to a stop in Hungary and the Czech and Slovak Republics in 1994, while Bulgaria and Romania - and of course the former Yugoslavia - still face a considerable decline.

## 12.3 Consequences for International Trade

This section will focus on trade implications of the integration of the Eastern European RMD economies into the world economy and rely on the study of Fischer and Rammer (1993). The focus will be on international trade flows. Particular emphasis is laid on three questions:

- *First*, what will happen to the overall volume of trade of the RMD area under a realistic catch-up scenario with respect to income growth? (Volume of Trade Effect)
- *Second*, what will the direction of trade look like? (Direction of Trade Effect)
- *Third*, what will the product composition of trade look like? (Composition of Trade Effect)

The central focus of the study is the very long-run outcome in the economy in which the Eastern European economies will have completed a successful transition, become thoroughly integrated into the world economy and behave like developed market economies in several aspects, especially in income levels (catch-up scenario). It is important to stress that even if the transitions are successful, this scenario is relevant only in the longer run, say in 2015. In particular, even with very rapid growth rates, it would take many years for real incomes in these countries to reach the average level of EU countries.

In addition to long term effects we tried to predict shorter-run issues and the more likely outcomes (short-term horizon: 1995 and medium term horizon: 2000). Trade during the 1990s will be shaped by two major developments. The first of these developments is the progressive liberalization of these countries' trade regimes. A more gradual process of trade liberalization has already made substantial progress in Hungary. The other countries - still behind - are likely to move in the same direction. The second major development is the dismantling of the Comecon trading arrangements which started in 1991. The official dismantling put an end to the complex system of bilateral trading arrangements as well as to the implicit pattern of subsidization of manufactured goods at the expense of natural resources and energy. These trade relations will become decentralized and market oriented, and show no substantial difference from trade with Western countries.

The methodological approach for analysing the consequences for international trade in the RMD area is based on a simple, but robust model of bilateral trade flows (a gravity type trade model, Fischer and Johansson 1994) and the above mentioned catch-up scenario (Fischer and Rammer 1993). In the sequel, we present our estimates with respect to the impact on the RMD trade. Of course, there is a great range of uncertainty with respect to the likely developments in each of the countries of the RMD area.

### 12.3.1 Volume of Trade

What will happen to the overall volume of RMD trade as the Eastern European countries liberalize and integrate themselves into the world economy in general and to the European Economic Area in particular? We consider the catch-up scenario for real incomes in the Eastern European RMD countries as described in Fischer and Rammer (1993) and report estimated trade effects in Table 12.1 (export figures) and Table 12.2 (import figures) for the RMD area. Due to time and cost constraints, the RMD area had to be considered as a closed rather than an open system.

The (long term) estimates are naturally subject to a great range of uncertainty with respect to the likely developments in each of the countries of the RMD area. The numbers presented might end up wide of the mark in the long term. But we think they are useful nonetheless in helping to frame the issues and focus thinking on the potential impact.

The overall trade volumes will expand primarily as a result of a rise in living standards and output levels, and increase the openness of the Eastern European countries concerned. Because any increase in income will take time, the global trade impact will be spread over a number of years. The total trade volume is predicted to expand by an average annual growth rate of 2.5 % in the shorter term and 4.5 % in the longer term. The main message of Tables 12.1 and 12.2 is that Eastern European RMD countries will make a major impact on global trade volumes only if the countries' income rises appreciably and this might be expected only in the long run. Moreover, the estimated figures imply that East-West trade in the RMD area will decrease in the short run, followed by a modest increase until the end of this decade and a dramatic increase in the long run. The bilateral trade volume is estimated to grow by a factor of 2.73 in the long run compared with 1990 under the assumption that the EU has removed restrictions on Eastern Europe's imports and all the Eastern European RMD countries have become members of the European Economic Area.

### 12.3.2 Direction of Trade

As already mentioned, the collapse of Comecon trading arrangements and the increasing influence of market forces in the Eastern European countries is likely to lead to a re-orientation of these countries' trade away from each other and towards Western European markets.

Figure 12.3 illustrates the spatial configuration of bilateral trade flows in general and the spatial re-orientation of Eastern European trade patterns in particular for the short, medium and long run in relation to the initial state in 1990.

**Table 12.1.** Actual and predicted export figures for the RMD countries (in 1990 billions of US dollars)*

| Countries | Actual Export Figures** | Model Predictions | | | |
|---|---|---|---|---|---|
| | 1990 | Short Run 1995 | Medium Run 2000 | Long Run 2015 | Change 1990-2015 (in % per year) |
| **Countries in the Catchment Area of the Danube** | | | | | |
| West Germany | 176.7 | 202.6 | 250.0 | 424.0 | 3.6 |
| Austria | 25.9 | 29.1 | 37.9 | 90.9 | 5.2 |
| Czech and Slovak Republics | 4.7 | 3.4 | 9.0 | 63.4 | 11.0 |
| Hungary | 4.8 | 4.4 | 11.0 | 73.7 | 11.5 |
| (former) Yugoslavia | 6.5 | 2.6 | 3.4 | 49.2 | 8.4 |
| Romania | 1.9 | 1.1 | 1.4 | 22.1 | 10.3 |
| Bulgaria | 1.1 | 0.4 | 0.6 | 7.6 | 8.0 |
| **Countries in the Catchment Area of the Rhine** | | | | | |
| Switzerland | 27.6 | 30.8 | 37.9 | 62.9 | 3.3 |
| France | 69.0 | 80.9 | 98.8 | 154.2 | 3.3 |
| Belgium-Luxembourg | 67.9 | 80.0 | 96.6 | 141.7 | 3.0 |
| The Netherlands | 69.3 | 81.0 | 98.2 | 146.0 | 3.0 |
| **Southeast European Countries in the Wider Catchment Area of the Danube** | | | | | |
| Greece | 3.3 | 3.6 | 4.5 | 8.4 | 3.8 |
| Turkey | 5.0 | 6.2 | 7.6 | 22.3 | 6.2 |
| **Total** | 463.8 | 526.0 | 656.8 | 1,266.4 | 4.1 |

* without trade relations Romania and Bulgaria-Romania ** UN World Trade Data Bank
Source Fisher and Rammer 1993

**Table 12.2.** Actual and predicted import figures for the RMD countries (in 1990 billions of US dollars)*

| Countries | Actual Export Figures** | Model Predictions | | | |
|---|---|---|---|---|---|
| | 1990 | Short Run 1995 | Medium Run 2000 | Long Run 2015 | Change 1990-2015 (in % per year) |
| **Countries in the Catchment Area of the Danube** | | | | | |
| West Germany | 121.1 | 138.3 | 172.5 | 305.0 | 3.8 |
| Austria | 31.5 | 36.3 | 45.8 | 91.7 | 4.4 |
| Czech and Slovak Republics | 4.9 | 3.6 | 10.1 | 66.7 | 11.0 |
| Hungary | 4.7 | 4.4 | 11.6 | 67.9 | 11.3 |
| (former) Yugoslavia | 9.9 | 4.2 | 5.5 | 72.9 | 8.3 |
| Romania | 1.8 | 1.0 | 1.3 | 20.1 | 10.1 |
| Bulgaria | 1.4 | 0.7 | 0.9 | 10.0 | 8.2 |
| **Countries in the Catchment Area of the Rhine** | | | | | |
| Switzerland | 39.4 | 44.6 | 54.0 | 80.3 | 2.9 |
| France | 98.8 | 117.0 | 141.6 | 215.3 | 3.2 |
| Belgium-Luxembourg | 69.3 | 81.5 | 98.7 | 143.7 | 3.0 |
| The Netherlands | 64.9 | 75.9 | 92.1 | 136.8 | 3.0 |
| **Southeast European Countries in the Wider Catchment Area of the Danube** | | | | | |
| Greece | 8.6 | 9.7 | 11.9 | 20.4 | 3.5 |
| Turkey | 7.4 | 8.8 | 10.7 | 35.6 | 6.5 |
| **Total** | 463.8 | 526.0 | 656.8 | 1,266.4 | 4.1 |

* without trade relations Romania-Bulgaria and Bulgari-Romania ** UN World Trade Data Bank
Source Fisher and Rammer 1993

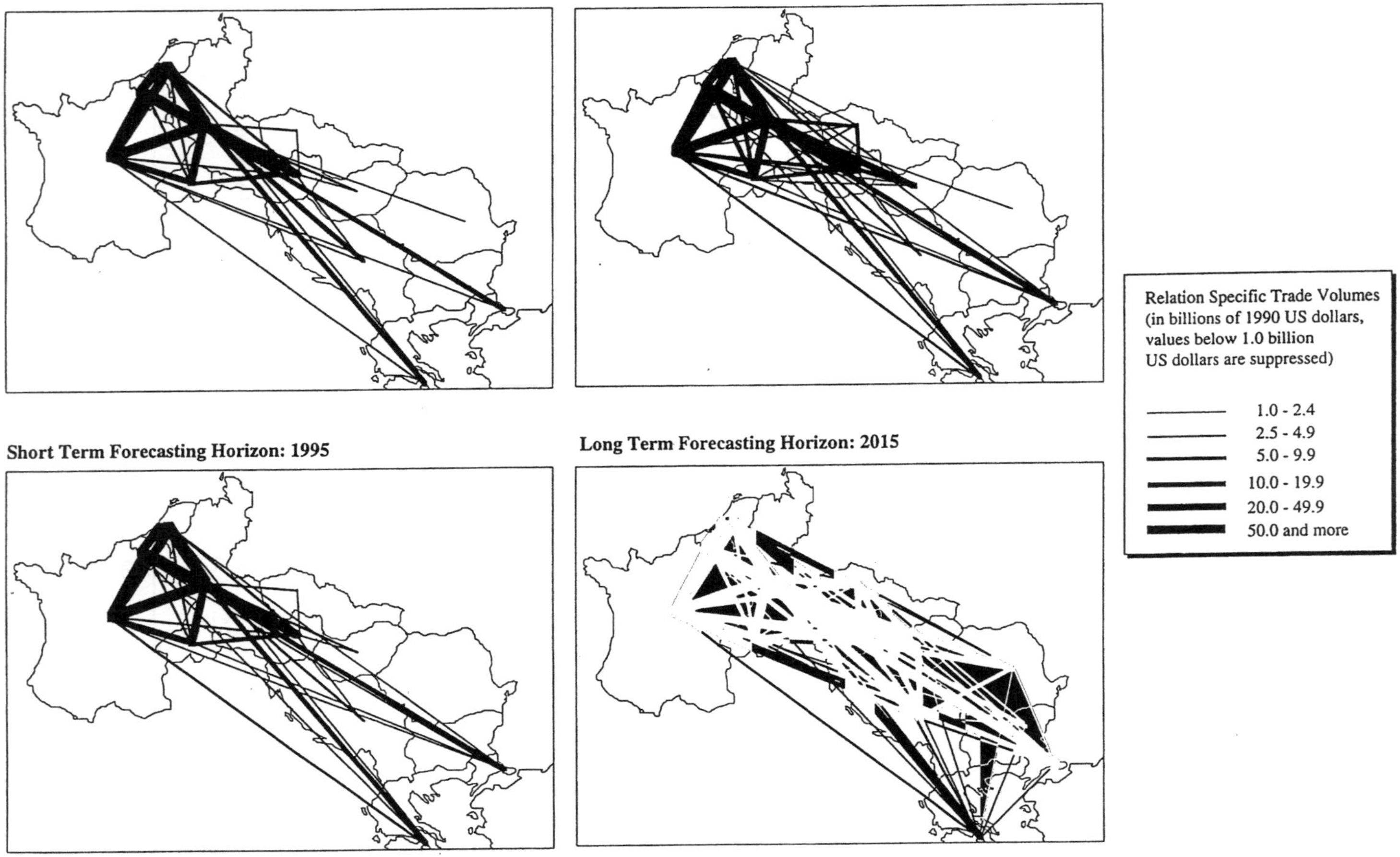


Source Fisher and Rammer 1993

**Fig. 12.3.** Spatial configuration of bilateral trade flows in the Rhine-Main-Danube trading area 1990, 1995, 2000 and 2015

### 12.3.3 Product Composition of Trade

So far the discussion has focused on volume and direction of trade effects. We briefly turn to the trends likely to emerge in the product composition of these trade flows. Two notes of caution are important to make in this respect. First, any projection based on current trade patterns faces the problem that this trade has been greatly distorted by centralised planning and misguided by domestic pricing policies in the socialist economies. Second, the likely expansion in the overall volume of these countries' trade, as described above, is unlikely to be a simple radial blow-up of today's. Significant trade expansion might involve a great deal of product diversification in exports.

In Fischer and Rammer (1993) six product classes have been distinguished to analyse changes in product composition of trade:

- Agricultural Products (SITC 00-09, 41-43)
- Raw Materials (SITC 21-25, 27-29, 32-33, 56)
- Labour Intensive Products (SITC 26, 61, 63-66, 69, 81-85, 89)
- Capital Intensive Products (SITC 11-12, 53, 55, 62, 67-68, 78)
- Low End R&D Intensive Products (SITC 51-52, 54, 58-59, 75-76)
- High End R&D Intensive Products (SITC 57, 71-74, 77, 79, 87-88)

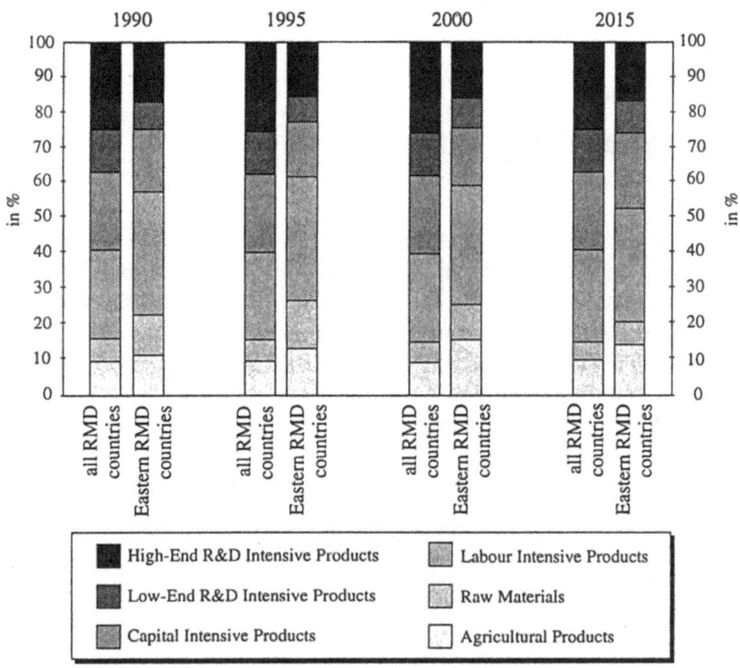


Source: Fischer and Rammer 1993

**Fig. 12.4.** Product composition in the RMD area: actual and predicted shares of product categories in total exports for all RMD countries and Eastern RMD countries

Figure 12.4 shows the product composition of trade for all the RMD countries on the one side and for the Eastern RMD countries on the other, in terms of the six above mentioned product categories (for country specific results see Fischer and Rammer 1993). The major aggregate results might be summarised as follows:

*First*, the comparative advantage patterns reflected in current East-West trade flows are undergoing major changes in the long run. Raw materials, in particular, which may loom large in the exports of Eastern RMD countries initially as well as in the shorter run, will be replaced in the longer run by manufactured products of increasing sophistication as their industrial sectors are rehabilitated.

*Second*, other product categories in which the Eastern RMD countries currently have strong revealed comparative advantages are agricultural and labour intensive products. The Eastern RMD countries are likely to develop a comparative advantage in goods that are intensive in human capital and skills, as the levels of educational attainment and scientific and technological mastery in these countries are high on average. The figures suggest that potentially disruptive export surpluses in agricultural products may well develop already in the medium term, if - as expected - the countries experience substantial productivity increases and if EU restrictions on trade (in agriculture) with Eastern European countries will be abolished.

*Third*, the Eastern RMD countries are likely to remain net importers of R&D intensive and capital intensive products. A large expansion of export in these product categories is unlikely although a slight increase might be expected in the long run.

## 12.4 Consequences for the RMD Waterway

We turn next to the implications and transport potentials for the RMD waterway. As Section 12.3 clearly indicates, the volume of trade between East and West in the RMD area will increase dramatically in the long run. In view of the missing links and bottlenecks in East-West transport on road and rail (Button 1993; Hall 1993), the RMD waterway link might play a major role in mastering the above mentioned challenge jointly with rail and road transport.

The estimations of transport potentials for the RMD waterway entail combining the predicted bilateral trade flows with levels of modal split to generate the freight potential for cross-country RMD waterway transport for 1995, 2000 and 2015. Therefore, bilateral trade flows - disaggregated by six product categories and measured in terms of US-$ - had to be transformed into freight potentials (measured in tons) disaggregated by the ten chapters of the standard NSTR classification system used in transport statistics (for details see Fischer and Rammer 1993).

We consider two scenarios for the modal split of the bilateral trade flows. The first scenario takes current modal split levels as given. This conservative scenario can be interpreted as a lower limit to the likely magnitude of trade potentials for the RMD waterway. The other scenario takes for granted that the RMD waterway transport is able to increase its competitive position. Its likelihood depends critically on overcoming bottlenecks in RMD waterway transport which mainly lie in:

- The fact that different parts of the RMD waterway network are state regulated or monopolised with different sets of rules and norms for modes of transport, type of cargo or container, type of investment, etc.,

- the lack of an intermodal uniform approach giving responsibilities for organisational issues to individuals (shipper-forwarder-receiver) rather than to governments (need for multimodal solutions),

- the lack of technical standardisation including physical infrastructures in port facilities and terminals for containerised transport as well as information systems,

- serious nautical problems on specific parts of the RMD waterway not guaranteeing an all-season minimum waterway depth of 2.5 m as well as small locks impeding larger vessels,

- the lack of support by transport policies of most RMD countries which are largely focussed on road and rail transport rather than on inland waterways, both financially and organisationally.

What can we say about the likely transport potential for the RMD waterway? Two sets of predictions of the transport potentials are reported. The first is based on the conservative modal split scenario, whereas the second assumes that the inland waterway will improve its position as described above. The results based on the conservative scenario suggest that the transportation potential of about 7 million tons in 1990 (for those traffic relations concerned) is likely to decrease rather dramatically to 4.9 million tons in the short run, increase modestly to a level of 7.9 million tons in the medium run and finally reach 32.7 million tons in the long run. At the other extreme, the optimistic scenario yields a transport potential of 13.0 million tons in the medium run and 72.4 million tons in the long run. While the optimistic scenario is a plausible scenario, we think its likelihood depends crucially on the implementation of the policy measures, improved nautical and logistic conditions etc. as mentioned above.

The expected transport potentials for the medium term are displayed in Figure 12.5 and Figure 12.6. The figures clearly show large disparities in the transport potentials between the Eastern and Western parts of the RMD waterway. In the medium run the potential for the Western part including the Hungarian harbours is expected to be much larger than that for the Eastern part. In the long run, however, the disparities tend to decrease if the economic transitions in the countries concerned become successful.

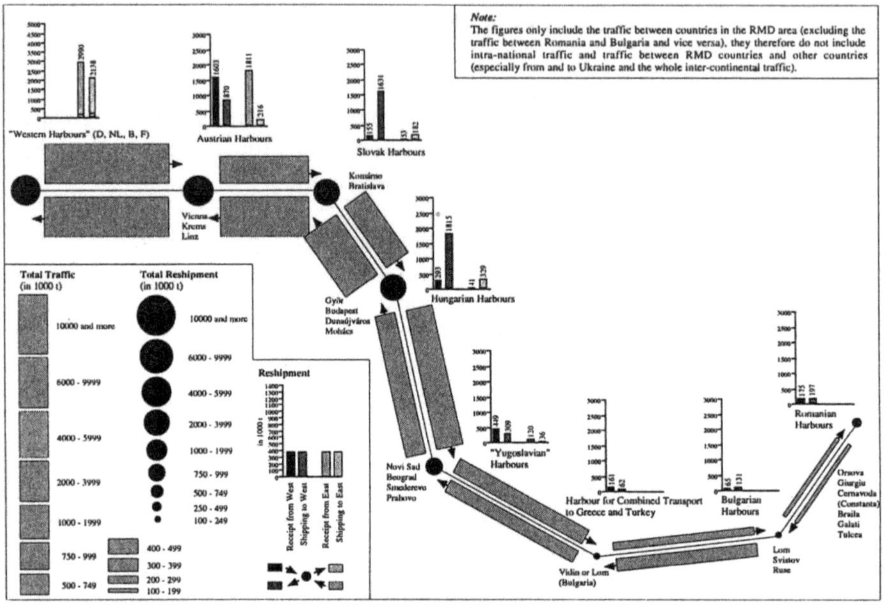


**Fig. 12.5.** Expected trade potentials for the RMD Waterway: time horizon 2000, conservative scenario

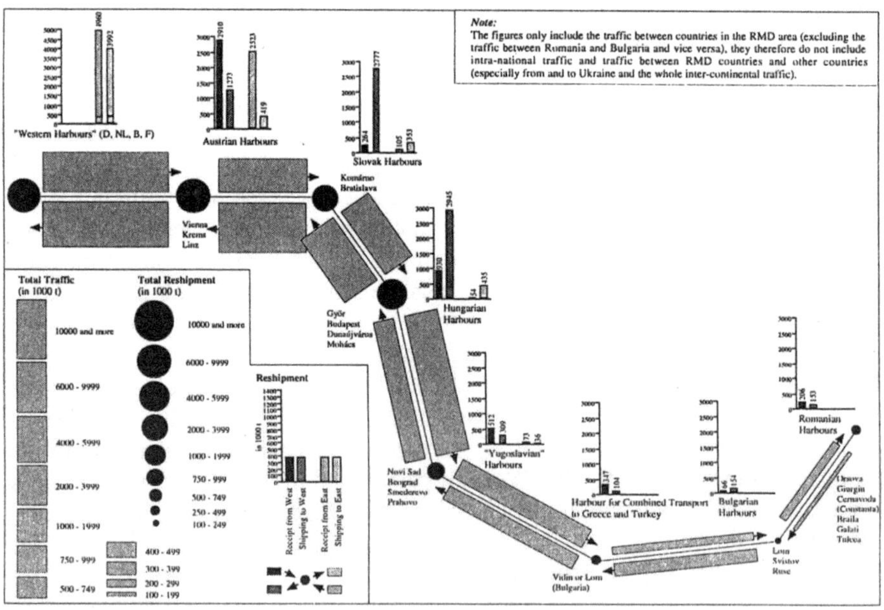


**Fig. 12.6.** Expected trade potentials for the RMD Waterway. Time Horizon 2000, optimistic scenario

## 12.5 Conclusions and Some Policy Recommendations

The peaceful revolutions in Eastern Europe are among the most dramatic events of the current epoch. A central element of the revolution in most of the countries is a total reversal of economic orientation. The major focus in this study was on the likely trade effects of the emerging market economies on RDM trade and the potentials for the RMD waterway transport.

Our trade predictions, based on a catch-up scenario for real incomes in the RMD countries, imply a sharp fall in East-West RMD trade in the short run, a modest growth in the second half of this decade and a dramatic increase in the first two decades of the next century. The estimated transport potentials for the RMD waterway are based on two alternative modal split scenarios (a conservative and an optimistic scenario). The results indicate an increase in transport potentials for the RMD waterway in the medium and long run while in the short run a decrease has to be expected.

The results on trade potentials clearly indicate a challenge for the RMD waterway in East-West trade. The realization of the potentials either based on the conservative or the optimistic modal split scenario crucially depends on five major dimensions (Fischer and Nijkamp 1993; Giaoutzi and Nijkamp 1993; European Round Table of Industrialists 1991):

### 12.5.1 Hardware

Hardware refers to the tangible physical components of inland waterway transportation infrastructure (e.g. technical equipment, ports, terminals). Bottlenecks include the slow substitution of outdated small ships by modern, larger ones implying overcapacity and growing inefficiency, the lack of standardization of dimensions of vessels and containers, and the lack of direct, multimodal lines between ports and hinterlands. Intermodal transfers are hampered by the lack of compatibility between barges, containers, port facilities and train terminals. Above all, the inland waterway network has to be based on standardized container technology. The quality of the existing equipment in most Eastern European ports is outdated. Ports and container terminals, heavy cargo and multiple bulk cargo terminals have to be equipped with advanced transhipment technologies which allow for a quick transmodal change of goods. Transhipment and cargo handling techniques have to be sought in both ship design, and loading and unloading, handling and storage operations.

### 12.5.2 Software

The term software is used here to include computerized management of transport and computerized services which improve access by users and facilitate intermodal operations (waterway/road, waterway/rail). Unsatisfactory statistical information on trade flows on the RMD waterway is a major issue. The second major bottleneck is the lack of informatics in general and EDI-systems in particular. Finally, the absence of logistic strategies as well as instruments of combined transport to control the wagon, truck and ship fleet on road, rail and waterway are other important shortcomings.

### 12.5.3 Institutional and Organizational Setting

The discussion encourages all regulatory, administrative, legal, management, and coordination activites and structures governing both the demand and supply side of transport. Market access limitations (licences), tariff setting procedures and ownership patterns are among the issues currently being discussed in the move towards liberalization, deregulation and privatization. Transport is a multimodal, multiactor and multinational activity which needs both competition and flexible regulation. In addition to traditional containerized transport solutions, the network has to be based on what is called soft technologies in combined transport. There is a strong need for more sophisticated and standardized equipment to be used in the terminals, in order to meet demand oriented criteria, inter alia reduction of transhipment time and personnel. In particular there is a need for integration, coordination and harmonization of regulations. Coordination will be achieved by making the different parts of the transport network compatible as a whole, including multi-modal solutions.

### 12.5.4 Financial Arrangements and Funding

Bottlenecks here include insufficient investment and planning of new infrastructure or upgrading of existing ones and for fleet modernization. Improvement of the RMD waterway network is hindered by a severe lack of coordinated European financing initiative/institutions in both the private and public spheres. A European approach to the integrated treatment of funding on the one hand and the equalisation of economic and environmental benefits and costs on the other is urgently needed. Certain resources required for coordination and organisation should be found via well regulated taxation systems.

### 12.5.5 Environmental Impacts

Certain rules similar to the MARPOL (the international treaty to prevent massive pollution) should also apply to environmentally dangerous transport behaviour in the inland waterway network and to harmonisation of regulations for environmental protection among the various parties involved in the network. They should also include a ban on unsafe ships.

The overall conclusion from the present chapter is that the RMD waterway offers a great potential to meet the challenge of large increase of East-West trade in the long run. Therefore, however, sophisticated hardware and software equipment is needed, sufficient financial resources have to be reserved for more strategic investments, strict eco-policies should be imposed and, finally, transnational coordination of policies on European network connections should be pursued.

# References


Button K (1993) East-West European transport. An overview. In: Banister D, Berechman J (Eds) Transport in a Unified Europe. Policies and Challenges. Amsterdam, North-Holland 291-313

Collins SM, Rodrik D (1991) Eastern Europe and the Soviet Union in the World Economy. Institute for International Economics, Washington

Debs RA, Shapiro H, Taylor C (1991) Financing Eastern Europe. A Study Group Report. Group of Thirty, Washington

Dornbusch R (1991) Strategies and priorities for reform. In: Marer P, Zecchini S (Eds) The Transition to a Market Economy. Volume I, The Broad Issues. OECD, Paris 169-183

European Round Table of Industrialists (1991) Missing Networks. A European Challenge - Proposals for the Renewal of Europe's Infrastructure. European Round Table of Industrialists, Brussels

Fischer MM, Johansson B (1994) Opening Up International Trade in Eastern European Countries. Consequences for Aggregate Trade Flows in the Rhine-Main-Danube Area. WSG Discussion Papers 40, Department of Economic and Social Geography, Vienna University of Economics and Business Administration

Fischer MM, Nijkamp P (1993) Inland Waterway Networks in Europe. The Neglected Potential, Bottlenecks and Policy Initiatives. WSG Discussion Papers 28, Department of Economic and Social Geography, Vienna University of Economics and Business Administration

Fischer MM, Rammer C (1993) Zur Entwicklung des grenzüberschreitenden Warenverkehrs im Einzugsbereich der RMD-Wasserstraße und der Mengenpotentiale der Binnenschiffahrt. WSG Research Reports 4, Department of Economic and Social Geography, Vienna University of Economics and Business Administration

Fischer S, Gelb A (1991) Issues in socialist economy reform. In Marer P, Zecchini S (Eds) The Transition to a Market Economy. Volume I. The Broad Issues. OECD, Paris 184-201

Funke N (1993) Timing and sequencing of reforms: Competing views and the role of credibility. Kyklos 46:337-362

Genberg H (1992) Transition processes in Eastern Europe. The sequencing of reforms. In: Eberwein WD (Ed) Transformation Processes in Eastern Europe. Perspectives from the Modelling Laboratory, Peter Lang (= Empirische und methodologische Beiträge zur Sozialwissenschaft 10) Frankfurt 63-94

Giaoutzi M, Nijkamp P (1993) Waterways as an alternative mode. In: Banister D, Berechman J (Eds) Transport in a Unified Europe. Policies and Challenges. Amsterdam, North-Holland 291-313

Hall D (1993) Impacts of economic and political transition on the transport geography of Central and Eastern Europe. Journal of Transport Geography 1:20-35

Hamilton CB, Winters LA (1992) Opening up international trade in Eastern Europe. Economic Policy 7:78-116

Hinds M (1991) Issues in the introduction of market forces in Eastern European socialist economies. In: Commander S (Ed) Managing Inflation in Socialist Economies in Transition, The World Bank, Washington 121-154

Rollo JMC, Stern J (1992) Growth and trade prospects for Central and Eastern Europe. World Economy 15:645-668

Siebert H (1991) The Transformation of Eastern Europe. Kiel Discussion Papers 163, Institut für Weltwirtschaft, University of Kiel

WIIW (Vienna Institue for Comparative Economic Studies) (1994) Leichte Erholung folgt einer schweren Rezession. Die Wirtschaft der Oststaaten 1993/94. WIFO-Monatsberichte 67:288-304

# 13 Prospects for the Conventional Passenger/Car Ferry in the Aegean

H. N. Psaraftis

## 13.1 Introduction: Scope of This Paper

A major category of ships operating in the Greek shortsea shipping system today is the fleet of conventional ferries for the transport of passengers, private cars, buses, motorcycles and freight-carrying trucks. These ships operate on a network of routes set up and approved by the Ministry of Merchant Marine. They have virtually displaced the traditional passenger-only coastal ships that provided service to the islands in the 50's and 60's. The latter ships still exist, but their numbers are steadily declining. Having lost a significant share of their market to ferries on long-haul routes, these ships seem to be losing the battle on shorter routes too (their main theater of operation today), this time to high-speed vessels, such as hydrofoils, catamarans, surface effect ships, etc.

In this paper we examine the role of the conventional passenger/car ferry in the Aegean (and in Greece in general) and discuss its prospects for the future. In addition to the 'standard' set of factors that may enter such a discussion (e.g., economic growth, infrastructure improvement, role of new technology ships, market deregulation, etc.), the paper also examines some factors that have received little attention so far but may very well be crucial to the actual picture the system may have in the future. We also discuss some scenarios that would directly impact on the operation of the ferry, as we know it today, identify the risks that this sector is facing and offer some recommendations on strategies and operational practices that should be followed in order to increase the efficiency of the present system.

## 13.2 Status Quo

Greek coastal shipping involves the movement of passengers and freight within a complex system of mainland-to-island, island-to-island and mainland-to-mainland connections. Although the actual number of islands is on the order of several thousand, most of these islands are very small and uninhabited, and about 70 islands are officially considered important from an economic perspective[1]. The islands and the mainland are served by a total of 138 ports. By convention, 'mainland Greece' is considered to also include the Peloponnese and Euvoia. As thus defined, the mainland has 42 ports. The rest of the ports (96) are island ports. Crete is Greece's largest island, with 8 ports.

An important concept in Greek shortsea shipping is what are known as the 'main lines' of the system. Each main line is defined by the Ministry of Merchant Marine as a set of geographical clusters that are internally linked by a network of ship routes (e.g., Piraeus-Crete, Piraeus-Cyclades-Dodecanese-Crete, etc.). There are a total of 31 main lines, spanning the entire system. There can be more than one main line serving a specific geographical area (e.g. for the Cyclades there are 10 different main lines). Within a specific main line, a variety of individual routes can serve the ports that belong to that main line.

If one excludes (a) deepsea cargo vessels that make direct calls to the mainland or the Greek islands from overseas destinations, (b) feeder ships that carry cargo either in bulk or as general cargo between the islands and the mainland, and (c) cruise ships that exclusively cater to the tourist industry, there are just a few categories of ships that operate in the Greek coastal shipping system on a regular basis.

Perhaps the most important category is the fleet of ferries for the transport of passengers, private cars, buses, motorcycles, and freight-carrying trucks. These ships operate on the 'main line' system described earlier. They have virtually displaced the traditional passenger-only coastal ships that provided service to the islands in the 50's and 60's. Such ships still exist (this is the third ship category), but their numbers are steadily declining. They also operate on the 'main line' system described earlier. Having lost a significant share of their market to ferries on long-haul routes, these ships seem to be losing the battle on shorter routes too (their main theater of operation today), this time to high-speed vessels, such as hydrofoils, catamarans, etc.

This evolution from traditional passenger-only ships to (mixed passenger/car) ferries is the first significant transformation of the nature of shortsea shipping in Greece in recent years. This transformation has developed over the course of the last 20-30 years and has been spurred on to a significant extent by rapid island economic growth and by significant infrastructure improvements in island ports in the late 60's and early 70's. This allowed for the first time ferry service to these islands.

It should be mentioned here that up until that time it was impossible for passenger ships even to dock at most of the islands, people having to embark and disembark by using small boats that approached the ship while the latter was anchored in the middle of the harbor. The construction of docks and piers in most islands ended this mode of operation and opened the door to ferry services.

---

[1] All statistics of this paper are drawn from a computer database maintained by MARTEDEC S.A. in Piraeus and from the Statistical Service of the Ministry of Merchant Marine.

The ferry mode of operation, practically nonexistent in Greek coastal shipping a few decades ago (except for very short distances and for the island of Crete that had adequate port facilities), has steadily grown since the 70's, by extending service to virtually all Aegean and Ionian islands. Due to the introduction of larger and larger ferries (or 'super-ferries'), this mode has ballooned to explosive proportions in the last 3-5 years, and now constitutes an integral and the strongest component of the system.

From recent data of the Ministry of Merchant Marine, the number of conventional passenger/car ferries above 1,000 GRT runs today at 108. The average capacity is 4,966 GRT, with the capacity of the largest unit being about 38,000 GRT. The average age of the fleet is by no means young: For the above set of ships, the average age is about 25 years, most vessels being conversions of older designs.

There are also 3 ships that are pure vehicle carriers (ro/ro).

As all fares are fixed by the Government, ferry operators (as well as all shipping companies) cannot compete on price. Instead, they try to improve the quality of services offered. Indeed, in recent years this quality has been increasing, by the introduction of more comfortable accommodation, air conditioning, etc.

There is also another vessel category that has become increasingly important in recent years. It concerns the emergence of 'fast' (30-50 knots) vessels. This category of ships is worthy of investigation because it has the potential to radically transform the nature of the shortsea shipping industry in Greece in the years ahead, and, as such, may also affect the role of the conventional passenger/car ferry. These vessels are hydrofoils (operating in the system for the past 10-15 years), and newer fast types such as SES, catamarans, and SWATHs.

The growing interest in the potential of these vessels has induced a heated debate on their technical, service, and economic merits for shortsea routes in Greece. Adversaries in this debate are the owners of and prospective investors in such vessels and the owners of the conventional shortsea services (ferry and pure passenger), who are afraid of losing market share to the fast vessels. Arbitrator to the debate is the Ministry of Merchant Marine, tasked by national law to issue permits for the operation of vessels, to approve routes, and to set fares for every type of service rendered. However, owners of fast vessels often complain that the current excessive rigidity in regulation is an obstacle to the issuing of permits, and many times suspect that lobbying by the owners of conventional ships is the real reason for such difficulties. Some of the owners of these vessels complain that they are forced by the Ministry of Merchant Marine to operate on routes that eliminate many of their natural advantages, such as routes with many intermediate stops.

In addition to the above information on transport 'supply', one can also obtain some insights into the 'demand' side of the equation, as well as on the growth of that demand. The following statistics are important in this respect (data are drawn from the MARTEDEC database and cover the period 1985-1988):

- In 1988, total passengers carried by coastal shipping in Greece amounted to about 9.86 million, compared with 8.17 million for 1985. This corresponds to an average annual growth of about 6.9%. By contrast, the Martedec database estimates that for the same period the total passengers carried *by air* to and from the Greek islands fell from about 3.67 million in 1985 to about 3.18 million in 1988 (an average annual drop of about 4.5%).

- Passenger cars carried by ferries grew from about 571,000 in 1985 to about 762,000 in 1988 (an average annual growth of about 11.2%, or nearly double that of passenger traffic). This figure alone testifies to the explosive growth in the ferry business in recent years.

- The number of trucks carried by ferries also grew, but more slowly, from about 355,000 in 1985 to 385,000 in 1988 (an average annual growth of about 2.82%). The slower growth can be explained because truck traffic depends more on the overall economic/industrial growth of a region and less on tourism, which is more linked to passenger car traffic.

- The breakdown of passenger traffic among ship classes during the same period (1985-1988) is as follows. Carried by ferries *and* pure conventional passenger ships: from 6.95 million to 8.38 million (a 6.86% annual growth). Carried by 'fast' ships (hydrofoils only in that period): from 1.22 million to 1.48 million (a 7.1% annual growth).

- Passenger traffic generally exhibits a strong *seasonality.* Seasonality is less pronounced for truck traffic, which is more steady because of its dependence on general economic activity rather than on tourism.

- In many islands, freight traffic carried by trucks onboard ferries is today of a comparable (same order of magnitude) volume with freight carried by feeder ships. This is a radical change from the modus operandi of the past (when freight carried by ferries was virtually nonexistent). This is so in spite of the fact that the bulk of truck freight is carried as a *service byproduct* by ships that are predominantly in the *passenger* business (either for passengers alone, or for passengers with their private cars).

This last observation raises the following natural question:

Why should a significant portion of the freight to and from the Greek islands be carried by ships that have been designed mainly for the passenger business? Or, could the efficiency of the system be increased by redirecting truck freight to dedicated freight carriers (ro/ro ships) and freeing valuable ferry space for passengers and their cars?

A more daring question is: If a new design of 'fast' (30-50 knots) ferry enters the market in the future and attracts a large portion of the passenger *and passenger car* business, and, if truck freight eventually goes to dedicated ro/ro ships, will conventional ferries, as we know them today, be at risk?

We shall attempt to answer these questions in Section 3.

In spite of the radical changes that have occurred in the system over the past 20-30 years (at least in terms of kinds of ships that operate in it, and in terms of passenger and freight traffic), there has been little change in the design of the network that serves the system. 'Main lines' are still more or less the same, many still emanating radially from Piraeus, with little connectivity between islands that do not belong to the same main line.

This situation renders island-to-island movement of freight and passengers cumbersome if both the origin and the destination of the movement are located on different main lines. In such a case, the movement has to go through Piraeus, even if the two islands are adjacent. As an example, the distance between the islands of Paros and Sifnos in the Cyclades is less than 20 nautical miles, yet until recently the only service between these two islands was by a 20-meter boat (passenger only) that only sailed when the weather was good, and whose speed was less than 10 knots. [In the summer of 1992 this situation was alleviated by the introduction of a larger ship, the PAROS EXPRESS, which now links the two islands, but still not yet on a daily basis].

The explosive growth in traffic that was spurred on by general economic growth and by the modernization of island port facilities in the last 20-30 years has in turn rendered many of these facilities obsolete and in urgent need of upgrading.

As far as ferry traffic is concerned, the most pressing need seems to be the gross inadequacy of ports, not so much in terms of docks or piers, but rather in terms of road access and parking facilities. Such inadequacy makes loading and unloading of vehicles onto ferries a very painful operation, involving traffic jams on the streets in and around the port, and gridlock and confusion at quay side. This phenomenon is observed not only at busy ports such as Piraeus, but also at island ports. The latter are frequently overwhelmed by vehicle traffic, especially if two or more ships want to use the port facilities at about the same time, something that is very common during the summer.

For all the significant change that has taken place in the picture of *freight* transport, the logistics of the transport operation that involves the shipment of freight via truck on board ferries is particularly cumbersome and run on an 'ad hoc' basis. Most trucking companies are a one-truck concern, serving a specific island. Many trucks (the so-called 'private use' trucks) belong to merchants who operate businesses on the islands (e.g., a grocery store) and purchase their supplies from the mainland. Other trucks are 'public use' trucks, providing service to customers who want to ship goods to or from the islands. There is no published schedule, but delivery usually takes place within a few days. There is also no established tariff for shipments, and truck owners are essentially free to charge what they wish. Under these circumstances, the transport of goods is dictated mainly by individual factors, cannot be planned in advance and may in fact become very expensive for the shipper.

Needless to say, there is no use made of any EDI (electronic data interchange) scheme for the shipment of cargo or passengers. For instance, it is still impossible to purchase a ticket at a place other than the port of origin of a trip. This means that one cannot buy a return ticket, and cannot even reserve space on board a ship for the return leg, unless that reservation is physically made by somebody at the destination port. This situation, although palatable in the past when everything moved extremely slowly, has now been recognized to be completely unacceptable, and there is recent talk of setting up a computerized passenger reservation system similar to that used by the airline industry. Many people agree that the time for such a system is long overdue.

## 13.3 Future Prospects

Making predictions for the future of the conventional passenger/car ferry in the Aegean and in Greece is certainly not an easy thing to do, given the uncertainties that surround many of the important issues. Nevertheless, it is important to realize that some of the underlying parameters that will be critical determinants of the future state of the system are known in advance, and this renders the formulation of possible scenarios less difficult (although by no means trivial).

One of the parameters that is known fairly well in advance is surely the target of an effective European economic integration, expressed by the implementation of the Single European Market as of January 1, 1993, and further reinforced by the Maastricht accords.

The first such factor is the considerable pressure on Greece so that the transport deregulation implied by the European economic integration is also applied to Greek internal sea transport. If this happens, services could be offered by ships not flying the Greek flag, and/or not employing Greek seamen (a situation completely different from the current picture). Needless to say, such a prospect is not looked upon very favorably by Greek shipowners or Greek seafarer unions, and the Greek Government has traditionally supported their position vis-a-vis the EC, on the ground of excessive threat to 'strategic national interests'. As things stand today, it seems that the EC has granted Greece a reprieve until the year 2004 for this deregulation to take effect.

A second factor that will be important in the future comes from the Maastricht accords. Barring non-ratification (or substantial renegotiation) of the Maastricht treaty, Greece (along with Spain, Portugal, and Ireland) is likely to benefit from the so-called 'cohesion funds'. One of the uses of these funds is to enhance the economic development of regions located at the EC 'periphery', by (among other means) strengthening their links to regions closer to the center. If this happens, shortsea shipping in Greece, along with the vital sea transport link to and from Italy, will benefit considerably by drawing funding for necessary infrastructure improvements.

A third factor that is not directly related to maritime transport but will certainly have ramifications for coastal shipping in Greece is the increasing deregulation of *air travel* in Europe. If foreign airlines are allowed to serve Greek islands (either from Athens, or directly from European cities), some of the passenger traffic that now prefers or is forced to take the boat may shift to air transport. As mentioned earlier, competition from air transport already exists in Greece. However, a deregulation of the airline industry is expected to further increase such competition, irrespective of whether or not the coastal shipping market is still internally regulated. This 'external factor' will provide even more pressure for the system to become more efficient in the future.

As said earlier, and similar to international practice, there has been a steady evolution in the types of ships that sail the Greek seas, in terms of size, speed, and technology. A ship class that is not yet thoroughly tested, but for which there exists considerable interest is the category of 'fast' marine craft. The basic question is what is the future of these new ship technologies.

The answer to this question could very well be critical for the future of the entire shortsea shipping industry in Greece, because depending on how the answer goes, the picture of the system in the future could be dramatically different from what it is today (much in the same way as the current picture is very different from what it was in the past).

Of these types of ships, the most widely tested in Greek waters is the *hydrofoil* class, which has managed to carve a special niche in the market. Perhaps its greatest advantage is speed (about 30 knots) and reasonably reliable and comfortable service. A basic limitation of this class is its inability to carry vehicles. This effectively limits its clientele to people traveling without cars, the same people who might consider taking the airplane, if this were available and attractive. Of course, the hydrofoil- airplane competition is less serious than it might seem, because hydrofoils typically serve islands very near Piraeus that are not connected by air transport. Another limitation of hydrofoils is that they cannot function properly in heavy sea states. Based on this, we think it is unlikely that this class will be able to further change the picture of shortsea shipping in Greece in the future.

*Surface effect ships* (SES) have been very recently introduced in Greece, and now serve some islands in the Cyclades as pure passenger carriers. Another SES is now operating between Brindisi (Italy) and the island of Corfu in the Ionian Sea.

Another potential category of advanced marine vehicles for Greek coastal shipping is that of *catamarans*. The catamaran class has had in the past a chance to demonstrate its merits in practice in Greek seas. Besides an unsuccessful attempt in the past to operate a semi-submerged catamaran between Piraeus and Crete (severe seakeeping problems), there are today several cases of successful operation of high-speed catamarans in Greece, mainly in protected waters (Argosaronikos Bay). Also, some lower-speed catamarans have been introduced in the Dodecanese. Plans exist for larger passenger/car ferry catamarans from the mainland to Aegean Sea islands, and for wave-piercers on the route from Patras to Italy.

Thus, experience from catamarans is still limited, and mixed. Some people like their speed, while some other people dislike their 'bumpy ride' in rough weather. It seems that if their service continues to be limited to passengers only, then catamarans will be more of a threat to hydrofoils (and to airplanes) than to conventional ferries. If on the other hand designs that are able to also carry private cars are introduced, then the entire shortsea 'game' will certainly be played by completely different rules.

A scenario that might conceivably be threatening for the future of conventional ferries in Greece is the possible emergence of ship designs that combine high speed and superior seakeeping performance, with an ability to carry vehicles. If such a design proves economically viable in practice, it may attract a significant share of passengers traveling with their private cars, away from the conventional ferry market. If, in addition to that, truck freight is diverted from passenger ferries to specialized freight (ro/ro) ships (only two of such ships currently exist), then the conventional passenger ferry fleet will be on a shaky ground in the future.

Of course, the two 'ifs' of the previous paragraph are really 'big ifs'. Although the two are not directly related to one another, it is perfectly possible that developments in one of them will facilitate developments in the other. We briefly discuss each of them below.

Coming to the second 'if' first, the development of a ro/ro fleet exclusively carrying freight to and from the Greek islands has never been systematically investigated in theory or tested in practice. However, it is conceivable that such a system, particularly if it is properly logistically integrated, be viable economically. If so, a considerable portion of truck freight may prefer to travel on these specialized ro/ro ships instead of the passenger ferries on which it travels today.

With respect now to the first 'if', the question is this: how serious is the possibility of developing a viable 'fast' design *that can also carry passenger cars?* This is perhaps the most critical (and certainly not easy) question to answer.

Several attempts to answer this question are currently under way worldwide (e.g., Techno-Superliner project in Japan, SES and SWATH designs, etc). In addition, the work of Papanikolaou et al at NTUA has obtained some insights into that question. They have proposed the design of a SWATH passenger/car ferry, to operate between Piraeus and Crete, and have optimized this design with respect to size, hull form, horsepower, and seakeeping behaviour. They have also carried out an economic analysis of this vessel. The main results showed a Required Freight Rate of USD 11.5/passenger, and USD 50.1/car. According to the authors, these compare with USD 8.6/passenger (B-class ticket) and USD 37.5/car for a conventional ferry running at about half the speed on the same route.

It is fair to say that additional research and/or a real-life test of such a vessel (or of other 'fast' types) will probably be necessary in order to fully ascertain their practical merits for Greek waters. Until such evidence is available, the extent of risk that conventional ferries run from competition from these vessels will remain unknown. An NTUA project on the prospects of Greek coastal shipping that has recently started will attempt to answer many of these questions.

With the various kinds of pressure on the system towards a better efficiency and improved levels of service, it is only natural that this will ultimately translate into an improved design of the shortsea transport network and into improved infrastructures.

Regarding the network, even a partial deregulation of the market (that is, with no removal of cabotage privileges) will result in the establishment of new routes, with minimal interference from the Government. The concept of 'main lines', as it is known today, may become meaningless for the overall fleet, although each individual company may very well establish its own main lines (which will not necessarily be the same as those of its competitors).

It is still premature to predict what a deregulated shortsea shipping network will look like, but it is very likely that it will be very different from the one currently in place. In the United States, airline deregulation in the 80's radically transformed the air transport network from one with predominantly long-haul routes to a 'hub-and-spoke' system involving passenger transfers. This does not necessarily mean that the same will happen in Greek shortsea shipping (transfers are much more cumbersome when traveling by ship). However, one could very well envisage a system where main destinations ('hubs') are served directly from Piraeus, and remote destinations are served by boats that originate from these hubs.

Finally, as regards infrastructure improvement, this is considered absolutely essential for enhancing the efficiency and the service quality of the system. Not only ports, but also basic services such as passenger/car reservations, ticketing, communications, and all related logistical services should be completely overhauled. The establishment of 'integrated' services, either for passengers or for freight could take place. For instance, door-to-door shipment of packages maybe offered by a company that has integrated truck distribution with ro/ro ship operation.

## 13.4 Conclusions

Based on the previous analysis, we believe that a number of conclusions can be made. These can be summarized as follows:

- There has been a drastic change in the pattern of sea transport to and from the Aegean and other Greek islands over the past 20-30 years. This change involved the establishment of the (mixed passenger/car) ferry as the workhorse of the fleet and the gradual reduction of the role of the traditional pure passenger ships.

- Transport of freight by ferry to and from the Greek islands, virtually non-existent in the past, is now comparable in volume with that carried by conventional feeder cargo ships. This is so despite the fact that ferries are designed for and cater to mainly the passenger business.

- In spite of significant changes in ship technology and the pattern of trade, there has been little change in the structure of the network that serves the islands.

- The infrastructure of ports in both mainland Greece and the islands is in urgent need of improvement to cope with the volume of passenger and freight traffic.

- The regulatory structure under which the system operates is excessively rigid and is the cause of many of the problems. Deregulation of European Community maritime (as well as air) transport will bring about changes in the shortsea transport network design and operation.

- The increased emphasis on 'cohesion', as stipulated in the Maastricht accords, will enhance passenger and car traffic between mainland Greece and the islands.

We believe that the picture of coastal shipping in the Aegean and in Greece in general will be as different in the future as today's picture differs from the past. On the face of these changes, the conventional passenger/car ferry faces some serious challenges, both technological and institutional.

It is obvious that those ferry operators who cannot adapt to this new more dynamic and competitive environment will suffer. Those who can best adapt, who can take full competitive advantage of the opportunities, and who offer superior service at affordable prices will succeed and establish themselves as the main players in the new game.

This paper supports the general conjecture that the role of the conventional passenger/ car ferry in breaking down the isolation between mainland Greece and the islands can be very significant. However, such a role cannot be realized 'automatically'. In particular, it is certain that it will *not* be realized if things are left as they are today. If isolation is to be broken, it is clear that a lot of work has to be done on the part of everyone, so that the system can best adapt to the new and dynamic environment of the years ahead.

## Acknowledgments

Another version of this paper has been presented at the THALASSA '93 Ferry and Cruise Conference and Exhibition, March 30-April 1, 1993 (aboard the N. KAZANTZAKIS). Parts of both versions borrow from the following cited paper.

## References

Psaraftis H N, Papanikolaou AD (1992) Impact of new technologies on shortsea shipping in Greece. Presented at the First European Research Roundtable Conference on Shortsea Shipping, November 26-27, Technical University Delft, the Netherlands

# 14 Connectivity and Isolation in Transport Networks: A Policy Scenario Experiment for the Greek Island Economy

R. Zwier, P. Nijkamp, F. Hiemstra and K. van Montfort

## 14.1 Towards an Open Network Economy

In the past decade the spatial-industrial organization of modern economies tends increasingly towards a *multi-actor and multi-modal network configuration*, in which *free access and competition* are becoming dominant features (Cuadrado-Roura et al. 1994; Nijkamp 1994; Porter 1989). In recent policy documents of the European Union it is taken for granted that the development of the European continent is shaped through a complex interplay of a networked physical space (composed of an open multi-regional system) and a multi-layer socio-economic structure (featuring open access and free competition), witness also the plans for the T.E.N. (Trans-European Networks) system. The drive towards a network economy - as a catalyst for accelerated economic growth - has been favoured by drastic changes in telecommunication and information technology (Capello 1994), including the use of advanced logistics and electronic data transmission.

The emphasis on network infrastructure in Europe is also caused by the strategic argument that one of the primary objectives of the European Union is to achieve and strengthen social and economic cohesion within the Union. Consequently, isolated areas have to be connected to more centrally located and intermediate regions or nodal points in the European network. It is expected that through improved connectivity the locational potential and advantages of isolated areas will improve. It should be noticed in this context however, that network connectivity offers a window of locational opportunities - in the form of necessary development conditions, but that no guarantee exists for an accelerated regional growth (see for a thorough study on this topic Bruinsma 1994). Nevertheless, it is important that new initiatives for a higher performance of the European network be made.

At the same time awareness is growing that the European network system is still a *fragmented* system, which does not maximize its potential benefits by far. This has been convincingly demonstrated in recent discussions on missing links and missing networks in Europe (Nijkamp and Vleugel 1994). Apparently, many countries still have an individualistic behaviour based on self-interest. As a result, we observe many impediments to a maximization of the economic synergy (and value added) of European networks. Apart from financial and environmental constraints, many other bottlenecks still exist, such as protectionist behaviour, cabotage restrictions, protection of national carriers, lack of international network cooperation, and lack of a multi-modal transport strategy from a European perspective.

The consequence of the above scene is clear: many regions in Europe are not well connected and even much more isolated than necessary. Unnecessary network constraints are not in the interest of people for whom the regulations have been designed!

It is also evident from recent policy discussions that network policy should not only address the issue of *(de)regulation* of the *demand* side, but also the *morphology* (and hence *design*) of a network on the *supply* side. The main emphasis should therefore be on a maximization of socio-economic benefits through *network connectivity*. Connectivity does not only have a physical meaning, but also an immaterial meaning (e.g. access to informal information or linkage to organizational networks). This is closely linked to the notion of *externalities* of networks (Capello 1994).

Generally speaking, networks tend to become the vehicles through which competition (and related value added) is favoured, and may hence be conceived of as critical success factors in an open economy. This clearly has implications for network policy, which should focus in particular on: *cohesiveness, openness, complementary logistics, private initiatives, network standardisation* and *deregulation*. Only under these conditions may one expect that networks will fulfil their catalytic role in a competitive environment. This holds at the overall European level, and also at specific levels, such as metropolitan economies, island economies etc. Based on the previous observations, networks can be typified according to the:

- structure of the transport market (free competition, regulated market etc.)
- type of mode (road, rail, waterways, air etc.)
- geographical coverage (from local to global)
- quality of service (including scale and scope), and the tariff system
- sophistication of transportation technology (e.g., logistic platforms, telematics, information systems)
- structure of the network (e.g., hierarchy, hub-and-spokes etc.)
- territorial and modal policy competence on networks
- barriers to a full performance of networks (e.g., regulations, conflict of competence etc.)
- integration with telecommunication (EDI, e.g.).

In the next section we will specify the above considerations with a view on our scenario case study for the Greek islands.

## 14.2 Network Policy Analysis: A Blend of Forecasting Models and Scenario Experiments for the Greek Passenger Shipping System

*Isolation* is not only a matter of *physical morphology* of a spatial system, but is also determined by *market conditions* (e.g. degree of competition) and *policy regulations* (e.g., fare structures or access conditions) in a transport system. Thus the topological features of a network (e.g. origin-destination links) are dependent upon physical geography, market forces and policies. *Technological progress* in transport systems may play a catalytic role in changing the degree of isolation (or peripherality), as it is able to influence physical bottlenecks or impediments to spatial accessibility, while at the same time it has an impact on market operations and policy initiatives.

The evolution of network use rests on an interplay between *demand* and *supply*. The demand side is associated with spatial mobility needs which have to be fulfilled with reasonable conditions (time, fares, reliability, frequency, comfort, flexibility, safety, etc.). Historically, the transport sector has - for equity reasons - always been an economic activity featuring much government intervention (including subsidisation). Only in the past decade, a strong tendency towards deregulation has emerged, starting in the U.S. airline sector, but gradually spreading to all transport sectors. With some delay, this is also happening in Europe, where deregulation is also accompanied by decentralisation and privatisation (Nijkamp and Blaas 1993).

The central variable that reflects isolation of an area or of a group of people is *distance*. From a geographical perspective, isolation of a certain area can then be expressed as physical distance to the geographical centre. The degree of socio-economic isolation of a group can be considered as the difference between their income per capita and the overall average income (Biehl 1986). From a mobility perspective, the degree of isolation of a specific region can be seen as the difference between the socio-economic capability of inhabitants of that region to travel to other regions and the capability of inhabitants of other regions to reach the first region. From a spatial access point of view the degree of isolation can be described as the extra travel time or cost and inconveniences that are attached to travelling to or from a certain region, in comparison with other regions.

Over the last few years, a trend in European transport planning can be witnessed, namely a shift in emphasis from demand to supply side policy. This has led to a shift in emphasis from mobility to access. This implies that transport planning is now focused more on passenger inconveniences, in both financial terms and in terms of extra travel time and other obstacles. The European transport policy will definitely have consequences for all network operations: road transport, railways, waterways, airlines, and public transport. The main idea is that a (re)regulation of the transport sector will improve overall network efficiency and will hence also favour the more isolated regions in Europe.

A prominent example of such an isolated area is formed by the Greek island system, which includes a large number of dispersed islands of different size categories and degree of accessibility. An additional disadvantage of the Aegean archipelago in comparison with other European peripheral regions is the fact that the Aegean islands (14.7% of the Greek population in 1981 (Statistical Yearbook of Greece 1989)) are scattered over an area that is almost equal in size to that of the Greek mainland. This makes access an even more difficult issue. Greek passenger shipping is considered necessary for both communication and integration as well as development of the Greek island and coastal economy, so as to minimize the isolation effect, especially during the winter months. The Aegean is one of the poorest regions in both Greece and the EU. In the eight inhabited North Aegean islands the GNP per capita is 35% of the EU average, while the 42 South Aegean islands show a GNP per capita amounting to 46.5% of the EU average, a difference likely stemming from the difference in tourist attractiveness between the two areas.

The traditional network system here was the shipping system, which was fairly well developed, but which is increasingly suffering from its low speed which places many islands at a vulnerable competitive edge. Competition in the shipping sector is also severely restricted through protectionist tariff regulations and cabotage prohibitions. In the new EU policy regulations for the transport sector, these unnecessary rules would have to be removed, but for the time being the Greek island system has - by way of exception - been placed in a privileged protectionist system, so that socio-economic inefficiencies will temporarily continue to exist. EU institutions have until now not shown any interest of awareness regarding the need to include ferry transport as an issue in transport infrastructure policy. The inter-island transport does not belong to the general integrated transport market policy of the EU. However, ferry links function as basic elements providing access in many countries in the maritime periphery of Europe (Jacobsen and Kristiansen 1992).

Nevertheless, the Greek shipping system is bound to go through a phase of economic, logistic and organizational restructuring, partly as a result of technological development, partly as a result of market conditions (e.g., hub-and-spokes configurations), partly as a result of EU regulations and partly as a result of changes in external policy conditions (notably the relations with Turkey). In our analysis we will mainly focus now on the shipping system in the Aegean Sea, as the geographical position of the islands in the Aegean makes them particularly suitable for network operations and analysis.

First, we shall discuss the *Aegean degree of interinsular connectivity*. Theoretically, the degree of interinsular connectivity can be described as the extent to which islands are linked to other islands in space and time. It is determined by topological aspects (availability or existence of links between nodes, resulting in morphology or network configuration), correspondence (existence of through connections and waiting times between different connections), frequency of lines, fluctuation of the frequencies of links over time (per day, week, month, season or year), and reliability of service (depending e.g. on climate/weather conditions (seagoing capabilities), on political conditions (labour strikes and shutdowns depending on degree of unionization) and on compliance with the pre-advertised timetable). Compliance refers to the performance of the company itself (punctuality) and to the presence of sufficient capacity on each ferry departure. Travel time is determined by both speed and connectivity. Speed and connectivity both have a positive relation to travel time.

Now, in order to describe the elements of Aegean interinsular connectivity, we turn to the function of the Aegean passenger shipping system. Its function is to play a strategic role in strengthening social and economic cohesion (and thus to overcome isolation) within the EU through connecting the various peripheral (from both a Greek and an EU perspective) Aegean islands with one another, with the Greek mainland and with other EU regions. The Aegean shipping network finds itself on the edge of the transport and communication network planned within the national framework, and is thus somewhat peripheral from a Greek perspective. In addition, the Greek transport network is also located on the south western periphery of the EU. So, the Aegean has a *double* (or super)*peripheral* feature. This results in inadequate and unsuitable links with non-peripheral regions. The current network configuration shows a clear radial pattern, the hub being the port of Piraeus. In recent years, also Rafina has gained significance as the second biggest passenger port of the Attica region. Frequencies are fairly high on most routes to and from Piraeus, but low on the majority of other routes. A more liberal policy towards granting concessions for new lines has been followed by the Ministry of Merchant Marine, over the last few years. Also faster types of vessels have been introduced, leading to more hub-and-spokes elements in the network. Furthermore, a strong barrier element at the eastern edge of the network can be noticed, caused by the still tense political relations with Turkey. This means that many connections with Turkey have low frequencies (or are not served at all) and are suffering from high taxes. The unreliability of the Aegean shipping network is largely caused by Greek labour union strikes and by frequent bad weather conditions during winter time. Barriers to optimal transfer possibilities are created by insufficient road access and parking facilities in and around seaports and lack of computerization, especially in the central port of Piraeus, which also has a serious telecommunication deficiency (Psaraftis and Papanikolaou 1992). Opportunities for improving the shipping system might be offered through such features as Computerized Reservation Systems and brand loyalty programmes, which are already extensively used in air transport.

*Connectivity* for transit transport is reasonable for many trips via Piraeus, but certainly not for other routes. In island ports, transit opportunities are often very bad, especially during the low season. Very few islands have regular links with one another. Most Aegean islands have a shipping connection frequency of once (or less) a week. During the winter this can even be less due to bad weather conditions and/or lack of interest on the part of shipping companies. The seasonal fluctuation is one of the most important characteristics of Aegean transport and probably the one causing most concern. The existence of peaks destroys comfort, limits safety and causes delays, while during off-peak periods continuity of service is at stake.

Seasonality is defined as the fluctuation of demand and, to a lesser extent, of supply over time. Use of maritime transport is traditionally high during the summer and low during the winter, due to fluctuations in tourism. Accordingly, the physical structure of the network also varies significantly between summer and winter as does the quality of service. Regardless of season, the quality of service also varies between short routes (only basic facilities) and more distant routes (more luxurious accommodation). An additional disadvantage of seasonality is the inevitable waste of capacity caused by an imbalanced spatial spread of demand, since huge flows on certain routes can be witnessed in only one direction at certain times. This implies empty vessels on the return segment of the route.

An analysis of seasonality over the period 1985-1989 (based on Goulielmos and Lekakos 1992) shows that 50 per cent of demand is spread over 25 per cent of the time, namely the third quarter. An extra 15 per cent of total annual traffic takes place in June (8.3% of the year). So, 65 per cent of passenger demand takes place in 33.3 per cent of the year, resulting in a peak demand during one third of the year of nearly double the monthly average and nearly four times the demand during the winter. A perfect supply (no under- or overcapacity) would imply that frequency times capacity per vessel from June till September must then be four times the average winter capacity. Since the government has ordered a certain minimum frequency during the low season, the largest part of capacity reduction must then be realized through operating smaller vessels.

In the early 70s the first ferry service to the Greek islands started to operate. The ferry mode of operation has grown steadily ever since. Due to the introduction of larger and larger ferries, "this mode ... constitutes now an integral and the strongest component of the system. Compared to other European countries (Northern Europe, Italy, Channel) their average speed is low" (Psaraftis and Papanikolaou 1992). But their speed is gradually increasing. On a number of routes hydrofoils are increasingly being used. Around 1,000 ships are currently operated in Greece. The average age of the fleet rose from 14.5 years in 1981 to 21 years in 1990. The average ferry carries about 700 passengers, the maximum capacity being 2,300 passengers and an average speed of about 16 knots (Psaraftis and Papanikolaou 1992).

Various entry barriers have been created to protect the Greek passenger shipping market from competition, both domestic and foreign. These barriers require *strong regulation*, which occurs in the form of required licenses, certificates of approval and other conditions, such as flying the Greek flag, Greek ownership, maximum age of 20 years, legal obligation to provide service also during winter months and so on. Hence, the operational freedom of the shipping companies is limited. Lines that are not profitable, but must be served for equity reasons (e.g. to isolated islands), are subsidized. The Greek government decision-making process has turned out to become significantly bureaucratic. "Decisions are long delayed because of paper shuffling between Ministries in Athens." (Lloyd's List International, 1991). Also, market transparency is flawed, which deteriorates the *perception* of network connectivity by passengers. Applying current marketing theories to networks would suggest that *real* connectivity is not valued by the passengers, but their perception of it. This perception improves when cooperation is established between shipping enterprises to improve communication towards the public with respect to prices, schedules, through connections, etc.

Furthermore, the operation of the Greek market mechanism is also limited by restrictions imposed on the functioning of the price mechanism. An Advisory Committee on *prices* was set up in 1992 and organized hearings in which the users of the Aegean network did not participate. It appears that "fares are not affected either by demand or by cost or ... by competition" (Goulielmos and Lekakos 1992). It is also a plausible assumption that fares are *fixed* higher than those of the competitive equilibrium but slightly lower than the monopoly level. This pricing policy obviously does not consider any explicit welfare calculations (Goulielmos and Lekakos 1992). It yields a high producer surplus and also renders higher cost to the consumer. The consequent lack of incentives to reduce costs results in economic inefficiencies. Since competition until now cannot take place in terms of prices, other features gain significance in competition, such as the same (or close) time of departure for the same destinations, time of arrival, duration of a trip and level of comfort (Goulielmos and Lekakos 1992).

The main aim of our analysis now is to identify the *driving forces* of changes in and the consequences for the future development of the Aegean shipping system. This requires first of all a network model depicting all forces at work and their expected consequences. According to a systems view on the transport field, the main factors that determine the future development of the network can be subdivided into four (interrelated) themes, namely a subdivision into the *internal* 'behavioural' mechanism, the *external* environment (both the *socioeconomic* and *technological* context) and the *control system* (made up by policy measures) (Nijkamp et al. 1990). Consequently, the factors that we consider to be the most interesting and dominant ones are now turned from a conceptual level into an operational level. They will prompt significant change to the network and dramatically alter the pattern of passenger flows.

The first dominant political factor is the *European unification* process. It is expected that economic growth in the region will be stimulated by EU policies, since Greece and the Aegean in particular are peripheral regions. As a result of this unification process, the cabotage restrictions are foreseen to be lifted in the year 2004. From then on, competition from non-Greek shipping companies in Aegean waters will be allowed. This leads to better standards of service, lower costs and thus lower prices.

The second political factor is the state of *Turkish-Greek relations and eventual Turkish EU-membership*. In 1994 these relations are still operating on a low level, resulting in the imposition of high transport barriers and thus in high transport costs between the two countries. If political relations improve, economic relations are also likely to ameliorate. Additionally, numerous forms of travel restrictions will be relieved. This may then lead to an increase in Greek-Turkish traffic. More and cheaper shipping connections can then be established, inducing a higher overall network efficiency and increasing attractivity for passengers. The centre of gravity of the network will probably gradually move eastward.

The last policy factor is the proposal for a *design of one or more (bimodal) hub-and-spokes systems by Greek network operators* in order to improve attractivity of the network. This type of network has a large degree of connection density and accordingly a shorter average travel time. This is beneficial to both consumers and transport companies. The companies are then able to offer a more attractive product and operate more efficiently.

Finally, a dominant external factor is accentuated by *technological innovation* in the past decade(s). It plays an important role, since the introduction and application of new technology has resulted in faster, cheaper, more comfortable, more silent and less polluting ships. These more attractive features will have a positive impact on the demand for shipping transport. Figure 14.1 presents a schematic overview of the principal factors.

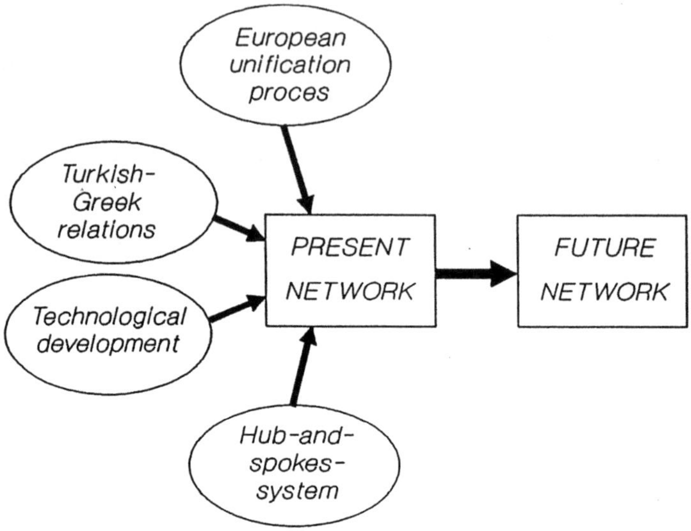


**Fig. 14.1.** Principal factors influencing the Aegean passenger shipping network

After the completion of the framework depicting the driving forces at work, a systematic assessment of long-range uncertainties has to be made, based on plausible or reasonable future scenarios for the area under study. Thus, we have to create an analytically feasible blend between predictive modelling and meaningful scenario developments. The differences between forecasting models and scenarios are illustrated in Table 14.1.

**Table 14.1.** Differences between forecasting and scenario analysis

| Forecasting models | Scenarios |
|---|---|
| □ Focus on quantified variables | □ Focus on qualitative pictures |
| □ More emphasis on detail | □ More emphasis on trends |
| □ Results determined by status quo | □ Results determined by future images |
| □ From present to future | □ From future to present |
| □ Deterministic analysis | □ Creative thinking |
| □ Closed future | □ Open future |
| □ Statistical-econometric tests | □ Plausible reasoning |
| □ From simple to complex | □ From complex to simple |
| □ From quantitative to qualitative | □ From qualitative to quantitative |

In our operational framework we will integrate scenario experiments with network modelling for the Aegean short sea shipping system. This means that a comprehensive network model has to be designed and calibrated first while next, scenario experiments are superimposed on this modelling system. For more background on short sea shipping, (see ECMT 1993; Wijnolst et al 1993; Gwilliam et al 1993).

We will characterize the demand side of the transport market in Section 14.3 and the supply side in Section 14.4. Then, Section 14.5 will discuss the design of a (deterministic) simulation model, coined ZKH[1], which models the evolution of the passenger shipping system. The model enables the use of quantitative impact methods. The input of the model consists of scenarios describing the way in which characteristics change over time or in what way they depend on other characteristics in each season. The output of the model is formed by the future values of the characteristics. The goal of the model is to modify in a clear way the development of basic characteristics and/or the relation between the characteristics and examine the effect these changes have on other characteristics of the transport market in the Aegean. Because of the close correspondence between these characteristics and the model variables, the characteristics may be conceived of the variables of the model. The various assumptions of the model will be given throughout the various sections. All assumptions made here were needed and have a background effect. In Hiemstra and Zwier (1994) a Decision Support System is presented as a software tool to work with the model[2]. In Section 14.6 we will define four scenarios which include the principal factors and discuss some analyses and results, corresponding to several scenarios. Finally, in Section 14.7 we will present some conclusions.

## 14.3 The Demand Side of the Transport Market

The demand side of the Aegean maritime transport system is described by means of a network. A network consists of nodes and edges. The nodes in the network of passenger shipping can either be a port on the main area or an island. A main area in our study is an area on the continent or a special island, which is selected for special reasons. We selected four main areas:

1. The Greater municipality of Athens (Athens Area), because it contains the most important port for transport to and from the Aegean, namely Piraeus.
2. The island of Paros in the Cyclades, because Paros may function as a potential future hub island.
3. The bigger municipal area of Izmir (Izmir Area), as it is important to analyze the consequences of vanishing political barriers with Turkey.

---

[1] ZKH stands for Zwier, Ko and Hiemstra. Zwier is the inventor, Ko symbolises all outside support and Hiemstra is the co-author.

[2] These authors also give an extensive discussion of the entire model and its constituents. For the sake of space limitation these details will not be given here.

4. The Turkish South-Westcoast (South West Turkey), because this area may also be used in analyzing the socio-political barriers between the nations.

The difference between the two Turkish nodes is that Izmir Area may be used more for transport between Athens and Turkey, while South-West Turkey has a significant touristic impact.

The other nodes are the islands in the Aegean. Figure 14.2 gives an overview of these islands, which we have selected according to their size. The smallest island is Arki, with a surface of 6 square kilometres. All islands (expressed in minors) and main areas (expressed in capitals) are represented by a point in a two-dimensional space. The nil point of this geographical map is 36.0 degrees Northern latitude and 23.2 degrees Eastern latitude.

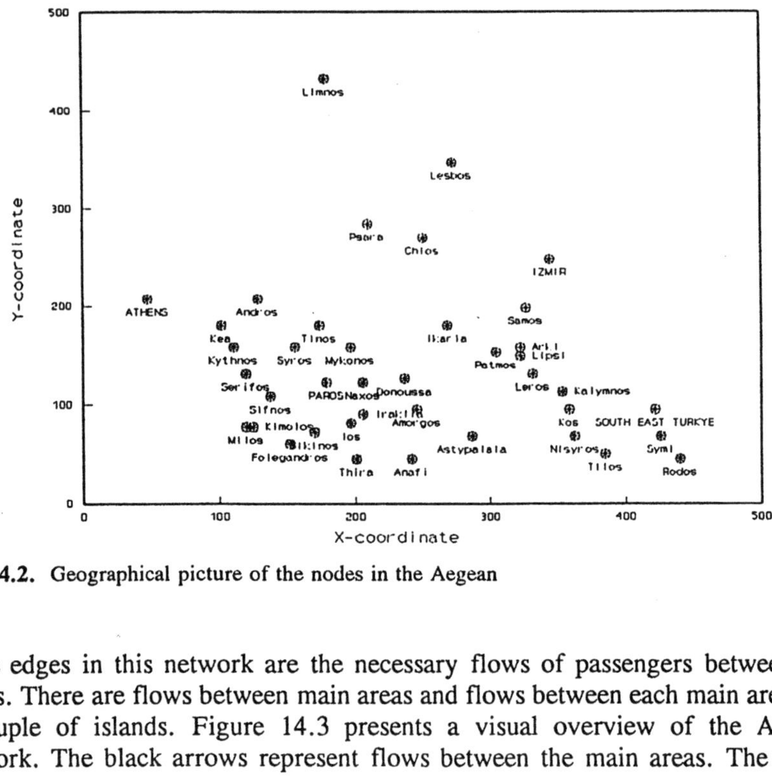


**Fig. 14.2.** Geographical picture of the nodes in the Aegean

The edges in this network are the necessary flows of passengers between the nodes. There are flows between main areas and flows between each main area and a couple of islands. Figure 14.3 presents a visual overview of the Aegean network. The black arrows represent flows between the main areas. The white arrows represent the flows between the main areas and the islands.

When a flow exists between one main area (or island) and another, it means that people wish to depart from a port in that area or on that island in order to go to a port in another area or on another island. An island is connected to a main area when there is a flow between these two. A flow from a main area to an island or between two main areas is assumed to be of the same size as its reversed flow. It is only important to know from where the passengers originate and where their destination is. Therefore, this network does not contain a specific flow structure constructed by the supply side. The supply side will be discussed in the next section. It is not necessary to know the exact geographical routes of the ships. Table 14.2 gives an overview of the characteristics we have selected to portray the network for the ZKH model.

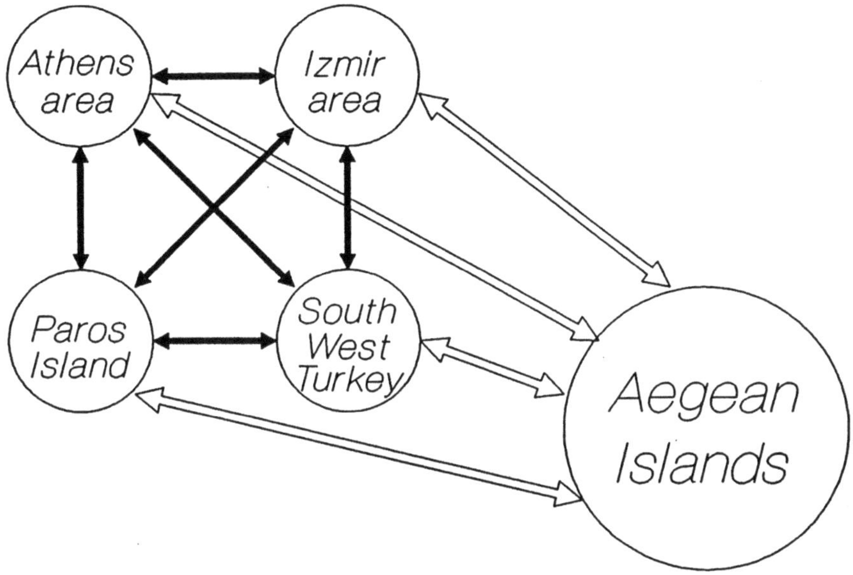


**Fig. 14.3.** The Aegean passenger shipping network

**Table 14.2.** Network characteristics of the Aegean maritime system

- The number of passengers that wishes to depart from a main area to go to another main area or island
- The number of islands connected with each main area
- The distance between each island and main area
- The distances between all main areas
- The ordinal ranking of connection priority of each island
- The connectivity of each island with each main area
- The relative attractivity of each island
- The demand for transport between the main areas
- The demand for transport between each island and main area
- The total demand for transport in the network

- The number of tourist accommodations on each island
- The number of inhabitants on each island
- The tourist image of the Aegean
- The port size of all ports in each main area that handle passenger shipping within the Aegean
- The existence of a political barrier with Turkey
- The geographical coordinates of the main areas
- The geographical coordinates of the islands

The rank order of connection priority reflects the actual spectrum of choice of people who wish to leave a main area in order to go to various islands. The connectivity deals with the number of main areas the travellers on an island wish to visit. All these characteristics will be translated into variables in the ZKH model. For the exact contents of these characteristics including their dimensions we refer to Section 14.5.

The Greek/Aegean travel pattern clearly shows a seasonal fluctuation. Therefore, we have subdivided a year into two seasons, namely a high and low season, or summer and winter respectively. The high season ranges from April 1st until the end of October. In our application we will start our simulation in the high season of 1993; this is season one. Thus, even seasons are low seasons and odd seasons are high ones. Characteristics cannot change within a season, but can only be changed over seasons.

## 14.4 The Supply Side of the Transport Market

The transport line is the set of business firms (companies) that provide passenger shipping transport in the Aegean and thus represent the supply side of the network. The vessels they have can be described by three types of characteristics (see Table 14.3).

**Table 14.3.** Vessel and network characteristics

| |
|---|
| - Technological |
| - Morphological |
| - Origin |

The main *technological* characteristic is the speed level of the vessels. There are two speed levels in our analysis. Speed level 1 indicates the set of slow boats and speed level 2 indicates the set of fast ships. A vessel is fast, when its average speed is higher than 40 km/h.

The *morphological* characteristic is the passenger capacity level of the vessels. There are two capacity levels. Capacity level 1 indicates small ships and capacity level 2 indicates large ships. A ship is large, when it has a minimal high season capacity of 200 passengers. A vessel with a certain performance is assumed to have only one particular average speed or capacity. We have decided to use these characteristics, in order to be able to analyze the competition between the modern hydrofoils (small and fast) and the traditional ferries (big and slow). The difference in speed level is nowadays approximately a factor 2 and the difference in capacity level is approximately a factor 10.

The *origin* characteristic is the flag of the ship, representing the country of registration. There are two types of flags: Greek and EU Non-Greek. Ships with a flag of the second type are registered in and owned by nationals of other EU countries. We use this characteristic to analyze the effect of the opening up and liberalization of the Aegean passenger transport market in the year 2004.

Altogether, we have defined 2x2x2=8 types of ships in the transport line. We assume for the sake of simplicity that every single company has only one type of boat, so that we can analyze the competition between the types of ships by creating eight distinct companies each of which represents a specific company in connection with a corresponding type of vessel. Thus the analysis of different types of ships is the same as the analysis of the competition between the companies. In order to visualize the transport line that consists of eight companies we have put them in a three-dimensional block (see Figure 14.4). The block has two storeys (flag differentiation), two side wings (capacity differentiation) and a front and back wing (speed differentiation).

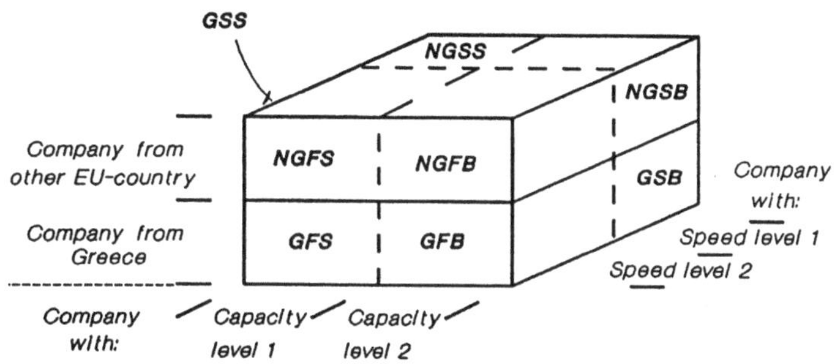


**Fig. 14.4.** Transport line block

The names of the companies are symbolised by three or four characters. The last character indicates the capacity level of the boats of the company: level 1 is S(mall) and level 2 is B(ig). The one but last character indicates the speed level: level 1 is S(low) and level 2 is F(ast). Finally, the first character(s) indicate(s) the flag of the boat: G(reek) or N(on) G(reek).

**Table 14.4.** Transport line characteristics

| |
|---|
| - The number of vessels of each company<br>- The price of each company<br>- The maximum volume of transport each company can produce<br>- The price each company asks relative to the other companies<br>- The relative attractivity of each company<br>- The average occupancy of the boats of each company<br>- The profit of each company<br>- The weighted speed level of all companies together<br>- The weighted price level of all companies together |
| - The price for transport set by the government<br>- The speed of the ships of each company<br>- The capacity of the ships of each company<br>- The fixed costs of each company<br>- The variable costs of each company<br>- The existence of a cabotage privilege for Greek companies |

The demand for and the supply of transport meet in the transport market. The market is subdivided into only two segments, namely into the segment of companies with ships with speed level 1 (slow boats) and a segment with ships with speed level 2 (fast boats). There is no segmentation into capacity level or flag, because we assume that passengers are not interested in the flag or capacity of the boat they travel with. Their only concern is assumed to be travel time and thus speed. Table 14.4 gives an overview of the characteristics that we have selected to model the transport line.

All these characteristics will be translated into variables in the ZKH model. This will be discussed in the next section.

## 14.5 The ZKH Model

The ZKH model contains all characteristics described in Sections 14.3 and 14.4. The characteristics have been subdivided into two groups. The first group reflects the endogenous variables and the second group reflects the exogenous variables of the model. An endogenous variable is a variable which is determined within the model by other variables. On the other hand, an exogenous variable depends only on variables outside the model or depends only on time. Therefore, the model characteristics can also be split into endogenous and exogenous ones.

Below, we will give an overview of all variables used. The variables denoted with an asterisk (*) are calculated in units of thousand within the model. The ranges of subscripts are $t \in \{1,\ldots,T\}$, $i \in \{1,\ldots,I\}$, $n,m \in \{1,\ldots,N\}$, $k \in \{1,\ldots,K\}$, $s \in \{1,\ldots,S\}$, $c \in \{1,\ldots,C\}$, with T being the number of seasons, N the number of main areas, I the number of islands, K the number of transport companies, S the number of speed levels and C the number of capacity levels. A variable with a subscript k is only used or determined, if the company operates with at least one boat within the transport line.

*Endogenous variables* (denoted in capitals) are:

| | |
|---|---|
| $AN_{n,t}$ | The number of islands that is connected with a port in main area n in season t |
| $B_{k,t}$ | The number of ships company k has in season t |
| $CA_{k,t}$ | The relative attractivity of a company according to the demand side of the transport market in season t |
| $D_{n,t}$ | The number of people per day that wish to leave a port in main area n to travel to another main area or island in season t (*) |
| $DI_{n,i,t}$ | The demand per day of people that wish to travel from main area n to island i in season t (*) |
| $DM_{n,m,t}$ | The demand per day of people that wish to travel from main area n to main area m in season t (*) |
| $G_{i,n,t}$ | The existence of a connection between main area n and island i in season t |
| $IA_{i,n,t}$ | The relative attractivity of island i according to the demand side of the transport market from the point of view of main area n in season t |
| $O_{k,t}$ | The average occupancy of the ships of company k in season t (full=100) |

| | |
|---|---|
| $P_{k,t}$: | The price in drachmas per passenger kilometre that a company asks in season t |
| $PK_{k,t}$ | The number of passenger kilometres company k produces per day in season t |
| $PL_t$ | The price level in drachmas of the transport line in season t |
| $PR_{k,t}$ | The profit per passenger kilometre in drachmas of company k per day in season t |
| $R_{i,n}$ | The ordinal number of an island for connection with a main area n |
| $RP_{k,t}$ | The price of company k relative to the price of the other companies in season t (x 100%) |
| $SL_t$ | The speed level in kilometres per hour of the transport line in season t |
| $SM_{n,m}$ | The distance between main area n and m |
| $SN_{n,i}$ | The distance between main area n and island i |
| $TS_{k,t}$ | The maximum supply of passenger kilometres per day of company k in season t |
| $TSD_t$ | The total demand for passenger kilometres per day in season t. |

*Exogenous variables* (denoted in minors) are:

| | |
|---|---|
| $bc_t^c$ | The passenger capacity of ships with capacity level c in season t |
| $bcc_{k,t}$ | The passenger capacity of ships of company k in season t |
| $bs_t^s$ | The average speed of ships with speed level s in kilometres per hour in season t |
| $bsc_{k,t}$ | The average speed of ships of company k in kilometres per hour in season t |
| $cp_t$ | The existence of a cabotage privilege only for Greek shipping companies in season t |
| $fc_t^{s,c}$ | The fixed costs in drachmas of a ship with speed level s and capacity level c |
| $fcc_{k,t}$ | The fixed costs in drachmas per ship per day of company k in season t |
| $gp_t$ | The basic price for transport per kilometre in drachmas which is set by the government in season t |
| $gpc_t^{s,c}$ | The price in drachmas set by the government for travelling with a ship with speed level s and capacity level c in season t |
| $ih_{i,t}$ | The number of inhabitants on island i in season t |
| $ps_{n,t}$ | The total size of all ports in main area n which handle passenger shipping in the Aegean (in hectares) in season t |
| $ta_{i,t}$ | The number of beds in touristic accommodations on island i in season t (*) |
| $tb_t$ | The existence of a political barrier with Turkey in season t |
| $ti_t$ | The tourist image of the Aegean in season t relative to the high season of 1993 (1993=100) |
| $vcc_{k,t}$ | The variable costs per passenger kilometre in drachmas of company k in season t |
| $xi_i$ | The x-coordinate of island i |

$xn_n$ The x-coordinate of main area n
$yi_i$ The y-coordinate of island i
$yn_n$ The y-coordinate of main area n.

The relation between the subscripts k and the corresponding companies and the relation between n and the corresponding main areas is given in Table 14.5.

**Table 14.5.** Subscript relations

| n | Main area |
|---|---|
| 1 | Athens area |
| 2 | Paros island |
| 3 | Izmir area |
| 4 | South-West Turkey |

| k | Company | k | Company |
|---|---|---|---|
| 1 | GSS | 5 | NGSS |
| 2 | GSB | 6 | NGSB |
| 3 | GFS | 7 | NGFS |
| 4 | GFB | 8 | NGFB |

The ZKH model contains 20 endogenous variables which are denoted in Figure 14.5 by ellipses and 15 exogenous variables which are denoted by boxes. The 19 variables concerning the demand side are positioned in the upper segment. These variables have a one-to-one relationship with the characteristics of the network shown in Table 14.2. The 20 variables concerning the supply side are placed in the lower segment and have a one-to-one relation with the characteristics given in Table 14.4. When an arrow leaves and enters the same ellipse, this means that earlier values of a variable are used in the determination of the present value.

A 'walk' through this model structure may offer more insight into the functioning of the model. Let us start at a certain value of the speed level(SL) and the price level(PL) of the system in a certain season. These variables - together with the port sizes(ps), the tourist image of Greece(ti) and the existence of a barrier with Turkey(tb) - determine the demand in the main areas for travelling within the Aegean(D). When we know this demand, we can calculate the number of islands(AN) people wish to travel to. Given this result and using the preference rank order(R) of the islands, we can figure out to which islands they wish to go (G). The exact flows between the main areas and the islands(DI) are determined by the attractivity of the islands(AI), which depends on the number of tourist accommodations(ta) and the number of inhabitants(ih). The flows between the main areas(DN) depend on the demand in both areas(D) and the distance between both areas(SN). On the basis of knowledge of the necessary flows(DI and DN) and the distances(SI and SN), we can calculate the total demand of passenger kilometres per day in the relevant season. This is the result of the first part of the model.

When we know the total result on the demand side, we may next assume that companies will compete in the normal Greek- or more commercialized European-way to get their share(PK), which is determined by the attractivity of each company(CA). The attractivity of a company is set by the kind of ships they have and the price they ask for. When the results, viz. (a) the profit(PR) depending on costs(vcc and fcc) and revenues(PK and P) and (b) the occupancy of the boats(O) depending on maximum and actual supply(TS and PK), are calculated, it is then time to set the price(P) for the subsequent season and to consider (dis-)investments in ships to get an new fleet size(B). The price setting depends on the existence of a cabotage privilege for non-Greek companies. Until the year 2004 the price will be set by the government. When the market becomes free, oligopolistic price policy will come to the fore. With the new prices(P) and fleet sizes(B) for the new season in mind, the price level(PL) and speed level(SL) can be determined and the next season is ready to enter the simulation model.

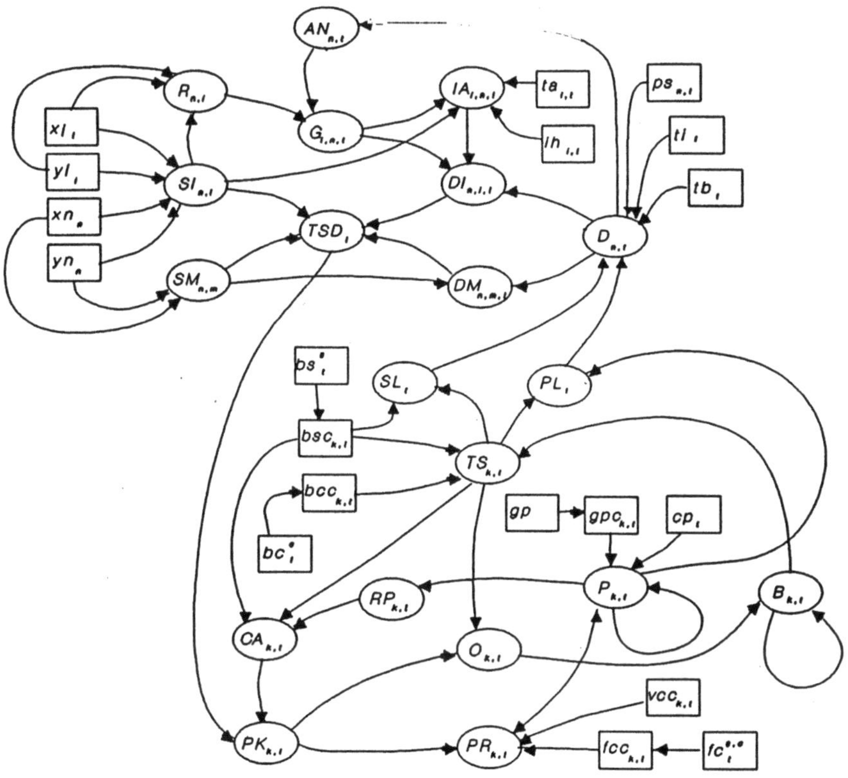


**Fig. 14.5.** Visualization of the entire ZKH model

The next interesting point is the way in which the principal factors are implemented in the model. The development of a hub-and-spokes system corresponds to a growth in port size of Paros($ps_{2,t}$). The development of the relationship with Turkey is built in via the existence of a barrier with Turkey($tb_t$) in season t. The technological innovation of the boats is implemented in the development of the speed of boats(bsc) and their fixed costs(fcc). Finally, the unification process is built in via the realization of the new prices and fleet sizes for each new season. The price and investment policy of the companies determine in what way profits(PR) and prices(P) affect the new prices and in what way the occupancy(O) affects the (dis-)investment in vessels.

In Hiemstra and Zwier (1994) the total model outlined above is split into 24 submodels in order to allow for a detailed analysis of the quantitative and qualitative relationships between the relevant variables. For example, one of the submodels deals with the realization of the flows of passenger demand between the main areas. A gravity model can be used in such cases, where the flow depends on a certain attractivity of the areas (here D) divided by the distances (here S). Thus higher attractivities will cause a higher flow of passenger demand between the areas, whereas on the contrary a longer distance causes a lower flow.

Two submodels which are rather interesting to take a look at are the submodels which realize the new price and fleet size for each company. The price setting is based on several price setting methods in a situation of an oligopolistic market. The methods deal with the price setting of the market leader and the immediate competitors, and the desired profit margin. The information given by the various methods is combined with fixed proportions.

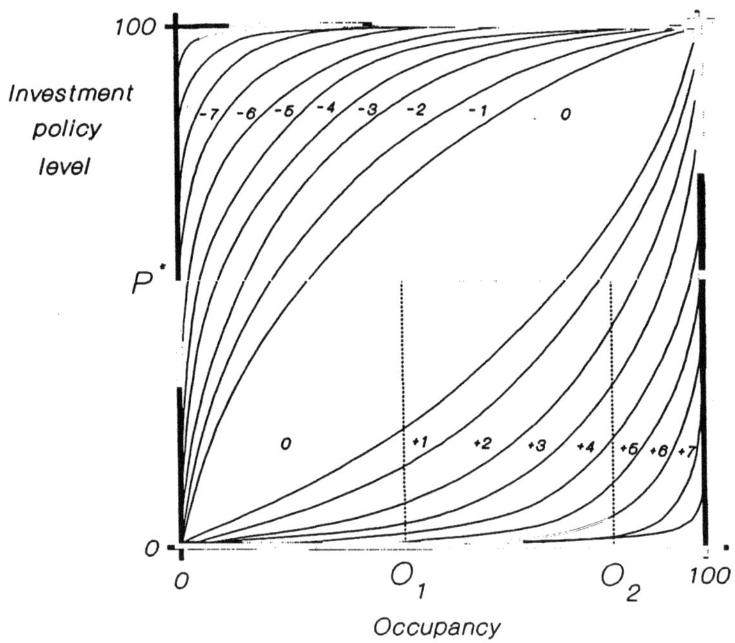


**Fig. 14.6.** Investment policy level in relation to occupancy

The investment behaviour and consequently the fleet size are based on the average occupancy of the ships in the last couple of seasons (see Figure 14.6, with the occupancy from prior periods on the x-axis and the investment policy level of the company on the y-axis). The figure is split into areas denoted by a plus or minus number, which indicates the number of ships in which a company should invest. A company chooses a policy level indicated by $p^*$. If the occupancy of the last high and low two seasons are $O_1$ and $O_2$, then these rates should be transferred to the policy line to find out in which area they are. For example, a value of $O_1$ which is in the '0'-area will generate no investment. On the other hand, if $O_2$ is in the '+2'-area, then the company will invest in two more ships. If $B_{k,t}$ becomes equal to zero, the company is out of business in the transport line.

The highest policy level is 100 and a less than full occupancy will cause the immediate withdrawal of ships. The lowest policy level is 0 and will cause an almost continuing investment in new vessels.

We have now discussed some submodels only in a qualitative sense, but without any quantitative representation. All variables are connected with one another in a way which is always open for revision and extension. In the next section some further quantitative results of the model we have built will be presented and analyzed.

## 14.6 Results and Analysis

To analyze the principal factors given in Section 14.2 we will now make use of the input of the ZKH model. The input information is generated by a scenario. A scenario is an abstract representation of the future. A scenario deals with a certain development of exogenous variables over time and with the realization of the price and fleet size in each season. We have created four scenarios and in Subsection 6.1 to 6.6 we will present the development of the variables PL, SL, TSD, D and PK over time under these scenarios. Furthermore, we will also present the results of the variables P, O, PR and B of the NGFS company over time. In Hiemstra and Zwier (1994) a Decision Support System is presented for easy scenario modification using the ZKH model. By using this software tool a wide variety of results can be generated, but we will stick here only to a presentation of a few interesting results.

The four scenarios are defined as follows. Scenario 1 is a basic scenario. The exogenous variables remain constant over time. All non-Greek companies start with 30 ships in 2004; all companies wish to have the same profit margin of 6 drachmas per passenger kilometre and try to set their price 5% below (underscore) the price of their immediate competitor. All companies have the same efficiency in constructing their lines; boats can operate during an average maximum time span of 12 hours a day. All companies have the same price and investment policy. The barrier with Turkey will remain; no hub-and-spokes system will be developed and no technological development is introduced.

Scenario 2 is equal to scenario 1 with the only modification that the barrier with Turkey will disappear in 2009 and Paros will be developed as a hub-island due to a constant growth of its port size over time.

Scenario 3 is equal to scenario 1 with the modification that the speed of fast ships will go through a constant technological development over time and that the competition between the companies after 2004 will become more intense. This implies that companies are satisfied with a lower profit margin, namely 2 drachmas per passenger kilometre, and that they try to set their price 10% under the price of their competitors. All companies will still adopt the same strategic and investment policy.

Scenario 4 is an extension of scenario 1, as it also incorporates the two previous modifications discussed above.

Subsections 14.6.1 to 14.6.6 only show high-season results, because no structural difference between the low and high season was noticed. Only a volume difference could be observed: the price level and speed level are equal in both seasons; the level of demand was 61% lower in the low season. The standard deviation was zero for all four high/low season ratios. The differences in prices set by the companies in both the high and low season did not significantly differ. Greek companies usually asked the same price, whereas non-Greek companies asked a price in the high season which was a little higher than in the low season, namely 2.3%. The occupancy for all companies in the low and high season had a ratio of 65 to 100, while the profit was quite equally distributed. The low season showed a slightly lower result compared with the high season, namely 96.9%. No big differences appeared in the fleet sizes in the low and high season, because a temporary low season withdrawal was not modelled.

### 14.6.1 Price Level Development Over Time

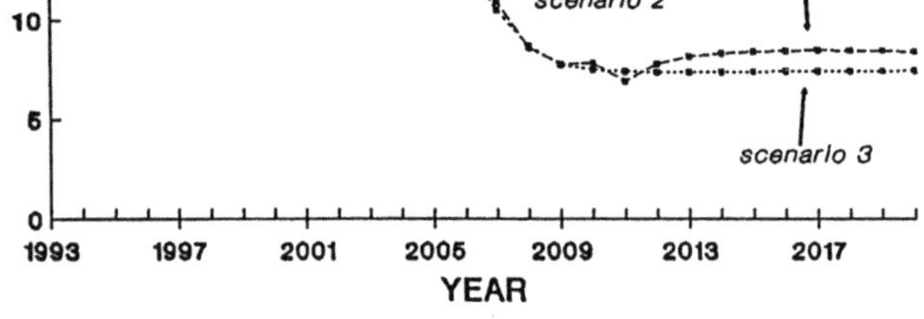


**Figure 14.7** Price level development over time

Until the year 2004 the price level development appears to be quite constant. See also Figure 14.7. Scenario three shows a small increase, because faster ships are coming up and may ask a higher price. When the market is set free, there is an immediate price drop. The fluctuation in the first two years is caused by the behaviour of the companies which try to get a feeling for price setting. Then a more constant decrease of the price appears. The little downward slope in scenario two is caused by the vanishing of the political barrier. Because of a sudden growth of demand, company results are favourable and the price can be set lower to get the desired profit margin, but finally a correction towards the old development is made. The price level in scenario three and four is lower because of the severe competition which was not assumed in scenario one and two.

### 14.6.2 Speed Level Development Over Time

Until the year 2004 the speed level seems to develop quite constantly, as is shown in Figure 14.8. The slight increase in scenario three and four is caused by the speed advantage which fast ships apparently have at the demand side of transport market, compared with slow ships. The non-Greek companies will enter the market especially with faster vessels. That is the reason why the speed level shows a quick increase. If the average speed of fast boats is held constant over time, then the market will be fully occupied by these types of boats in the year 2016. An increase in speed as illustrated in scenario three and four will cause the market to be fully occupied by fast ships in the year 2007. The small change in scenario two in 2012 is caused by the political barrier which by then has disappeared, so that an additional demand will apparently cause an extra input of more fast instead of slow ships.

### 14.6.3 Total Daily Demand Flow Development Over Time

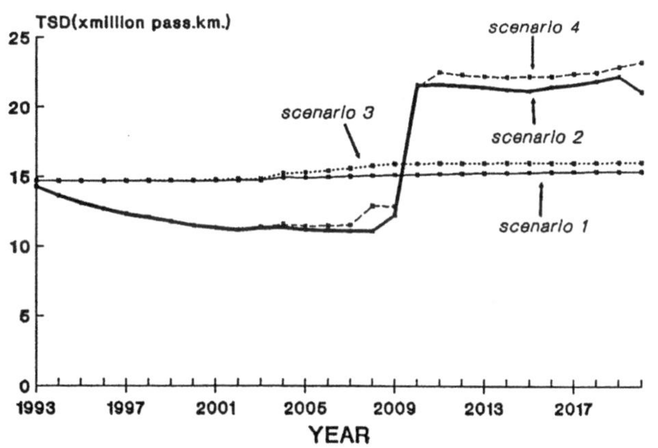


**Fig. 14.8.** Speed level development over time

In the high season of 1993 an average of 15 million passenger kilometres were demanded and produced per day. See Figure 14.9. It is obvious that under scenario one no big change is expected. Almost everything is held constant. The substantial changes are caused by the hub scenario and the barrier scenario. When Paros becomes extensively developed as a hub island, passengers will not have to travel via Athens anymore to arrive at another island. Travel distances are getting shorter. As can be seen in Figure 14.9, demand is constant until 2009. Thus, it might be expected that the total demand flow may decrease. The big crack in 2009 is caused by the vanishing of the Turkish political barrier which is implemented in scenario two and four. The daily demand flow does not drop any further because total demand increases.

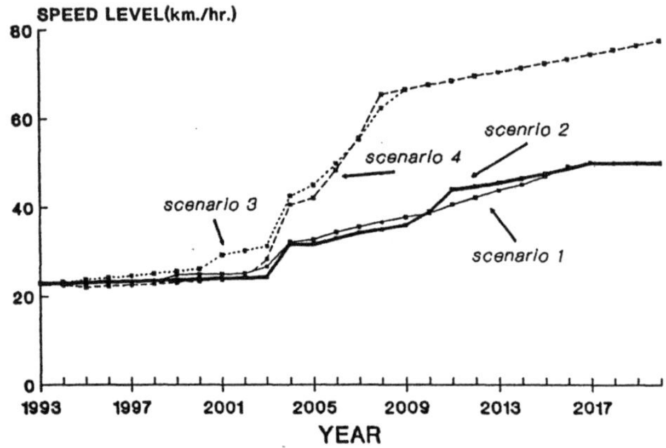


**Fig. 14.9.** Total demand flow development over time

### 14.6.4 Total Demand Per Day

It is noteworthy that the development of a hub island will not cause an increase in demand, only a geographical/topological redistribution. This is also demonstrated in Figure 14.10. Apparently passengers do care less about travel time. But the combination of the non-existence of a barrier and the existence of a hub will surprisingly cause an increase. This may be caused by the hub, which has a higher performance, when the demand in the eastern side of the network increases. In Figures 14.11 and 14.12 the geographical demand development has been depicted. We have visualized six demands flows, namely TI(the demand between Turkey and the island), PI(the demand between Paros and the islands), AI(the demand between Athens and the islands), PT(the demand between Paros and Turkey), AP(the demand between Athens and Paros) and AT(the demand between Athens and Turkey).

The demand composition of scenario three and four is similar to that of scenario one and two, respectively. As Figure 14.10 shows, scenario one did not show large total fluctuations. Also, the composition is quite stable. Thus technological development and a change in cabotage privilege appear not to prompt a change in the geographical distribution of demand. On the other hand, the previous discussed combination of scenarios causes - besides a change in volume - a change in composition. Of course, the demand flow between Turkey and the other nodes shows an immediate increase, but as a result the demand between the hub and Athens increases drastically. Although the total demand did not really fluctuate until 2009, the composition shows a certain substitution between Paros and Athens in the demand flow between these nodes and the islands. The constant increase of the total demand, as shown in Figure 14.10, can now be clarified by looking at their composition.

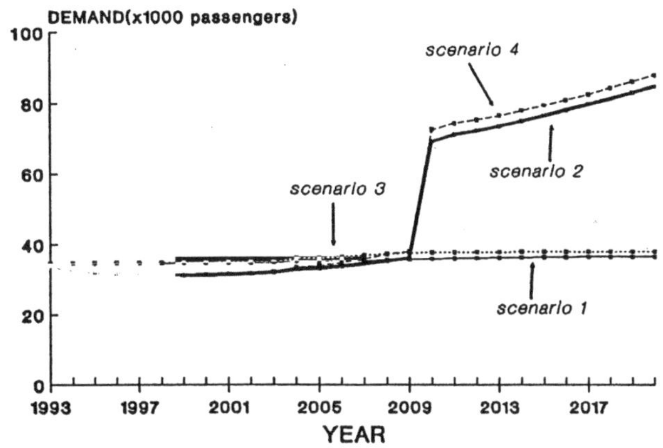


**Fig. 14.10.** Total demand development over time

### 14.6.5 Market Share

The distribution of passengers is at present quite stable, but with a free market and the technological development coming up, changes in market shares cannot be avoided. The situation in 1995 gives a clear overview of the market share of each company before 2004. The big and slow ships fill the transport line to a large extent. The big fast ships are highly underdeveloped and the hydrofoils (small and fast) have an attractive starting share.

There is much to say about each scenario in the high season in the year after the free market has started (2005) and in the high season of 2015. Therefore, we will first discuss scenario one and review the difference with the other scenarios later.

Scenario one shows a good start for fast and big vessels from non-Greek companies (36%) and a majority market share after 2015 (56%). See also Figure 14.13. Their Greek competitors have a positive market share development with a smaller growth rate. Although fast vessels seem to conquer the market, small ones seem to face some difficulty in obtaining more market power. The Greek company GFS seems not to be influenced by the free market. Its market share remains constant over time.

It could be expected that slow ships were losing ground. Also, as can be seen in the year 2015, the only slow ships that will still exist under scenario one in that year are the Greek traditional (big and slow) ferries. But they appear to be moving on their last engines. The fact that time is running out for slow vessels was already clearly shown in 2005 when the Greek big and small ships lost their market share compared with their initial situation. As could be seen before, passengers do not travel more when travel time decreases, but if they have to choose between fast and slow ships, then speed appears to be an important aspect in their decision-making process.

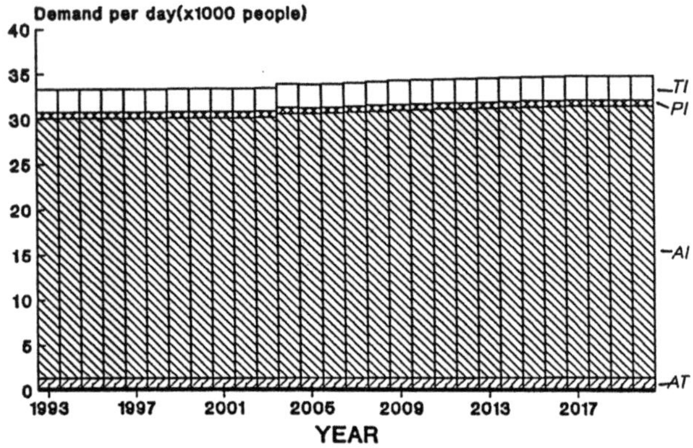


Fig. 14.11. Demand composition of scenario 1

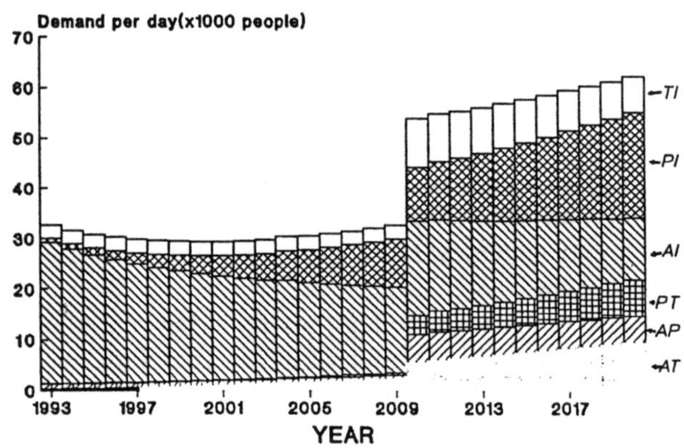


Fig. 14.12. Demand composition of scenario 2

Scenario two offers a more modest role for the big fast ships of Greece, as can be seen from Figure 14.14. This result stems from the hub which causes a decrease in the total demanded passenger kilometres. The NGFB company performs well in that situation. The reason that the GFB market share equals zero in the year 2015 is that we have modelled a no-return clause, once a company reaches zero capacity and is leaving the market. We assume that company NGFB takes over the market share that GFB could have in that case.

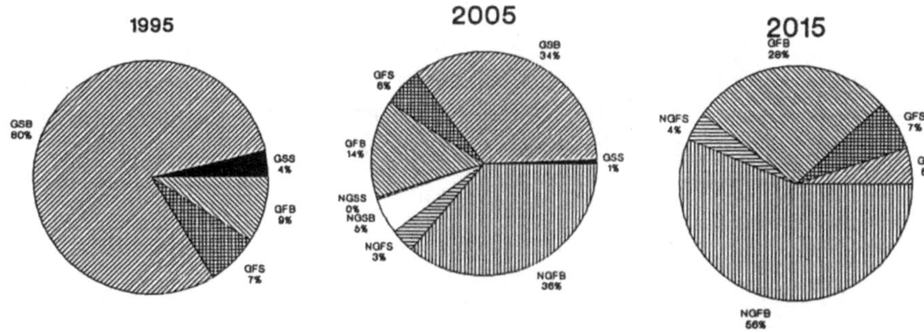


**Fig. 14.13.** Market share development under scenario 1

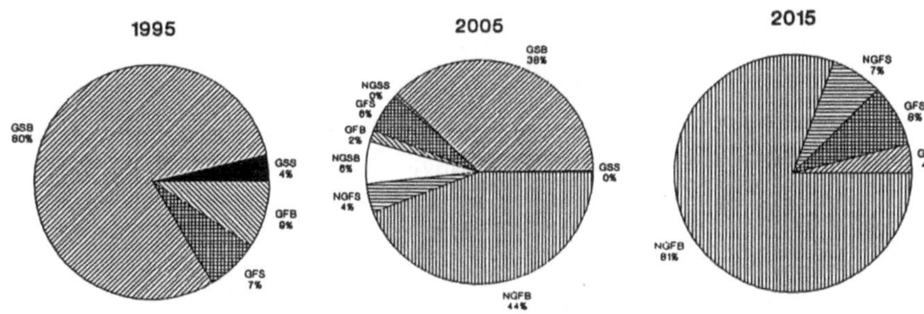


**Fig. 14.14.** Market share development under scenario 2

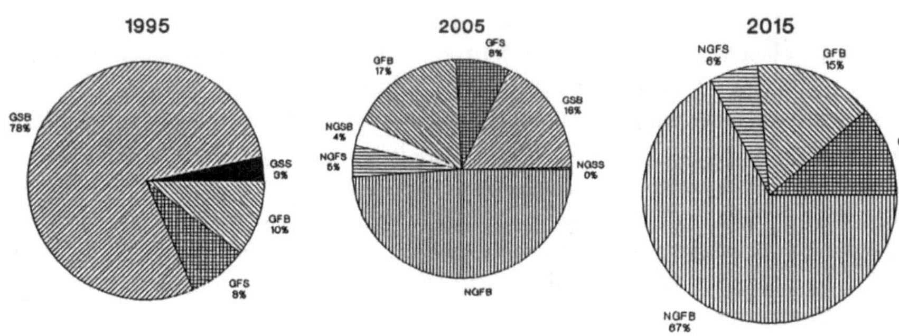


**Fig. 14.15.** Market share development under scenario 3

Scenario three shows a better emergence for the big fast ships of the NGFB company. See also Figure 14.15. A tighter competition leaves a 79% market share for companies with fast boats. Especially the Greek company with traditional big and slow boats seems to disinvest extremely: its market share decreases from 78% to 18% in 10 years.

Finally, in scenario four an ultimate victory of the non-Greek big fast boats appears in the year 2015. This can be seen in Figure 14.16. Greek companies have only 9% of the market in their hands. This is alarming for a country with such a strong historical shipping background.

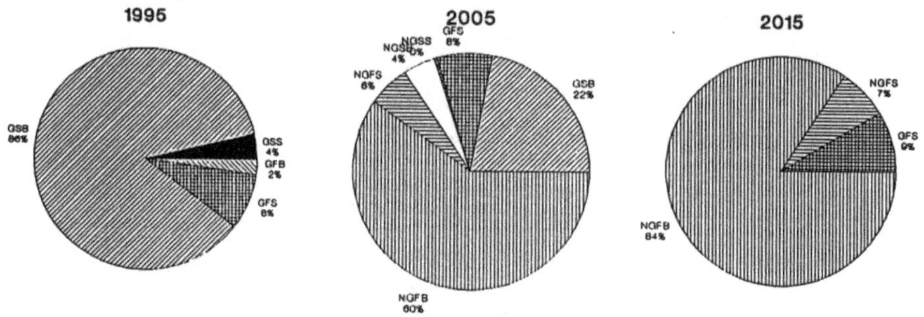


**Fig. 14.16.** Market share development under scenario 4

### 14.6.6 NGFS-Company

Now, we will look at the results of the NGFS-company. The price, occupancy, profit and fleet size will be compared at the same time. If Figure 14.7, the price level, is compared with Figure 14.17, then the price of NGFS has a close correlation. The free market causes a price drop in the first ten years. The slight slope in 2011 can now be clarified more clearly. It has to do with the disappearance of the political barrier. Because of the demand explosion, the occupancy gets a severe swing, which brings good results to the company, so that a lower price can be set to achieve the needed absolute profit margin.

The difference between the peaks in Figure 14.18 of the occupancy for scenarios two and four is caused by competition, which leaves a smaller market share for the company in scenario four. This difference in occupancy causes the difference in the development of the fleet size under scenario two and four. See Figure 14.20. The non-barrier advantage can clearly be seen in the higher fleet sizes that occur. The modest drop in the fleet size right after 2004 is caused by an initial input of ships which was too big.

It is interesting to see that the model generates a stable situation in the more distant future. Since however the demand still grows, the extra demand will not be allocated to the companies with small ships but to those with larger ships. The stability is primarily caused by a steady occupancy which induces a non-investment behaviour. A constant fleet size and a stable occupancy appear to produce a stable profit and, because the method of price setting according to the profit margin weighs heavily in the decision-making, the price also shows a balanced development. The influence of this price setting method is substantial. This method incorporated a six drachmas absolute profit margin in scenarios one and two, and a two drachmas profit margin in the scenarios three and four, and nevertheless a convergence appears towards these values six and two. Even if another price setting scenario were to have been chosen, it is plausible that there would be still a convergence, although the equilibrium values would be reached at a somewhat slower pace.

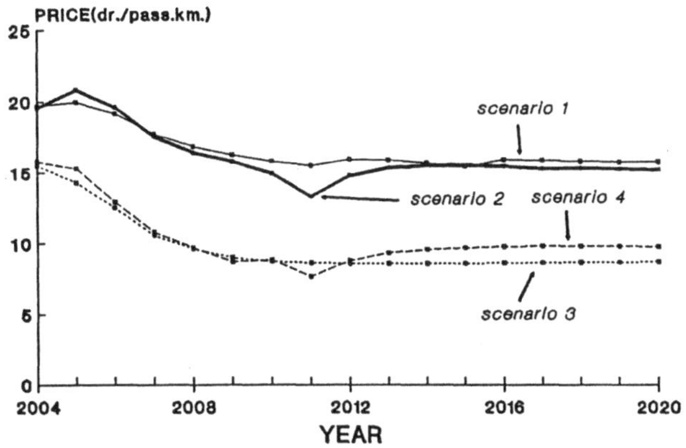


**Fig. 14.17.** Price development of NGFS

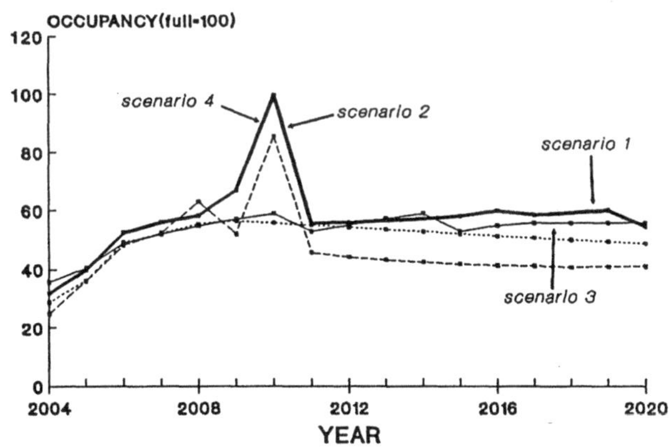


**Fig. 14.18.** Occupancy development of NGFS

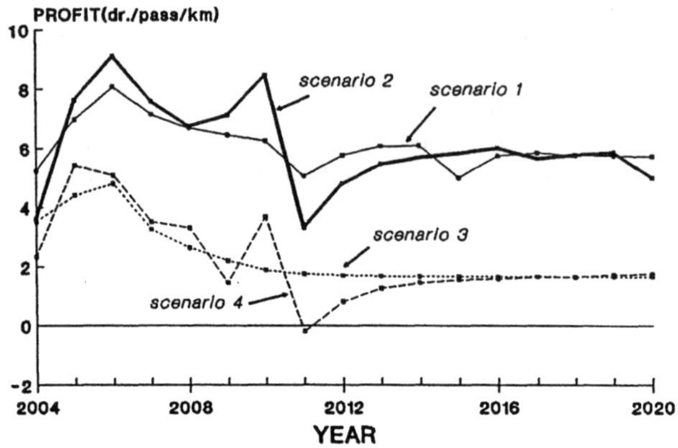


**Fig. 14.19.** Profit development of NGFS

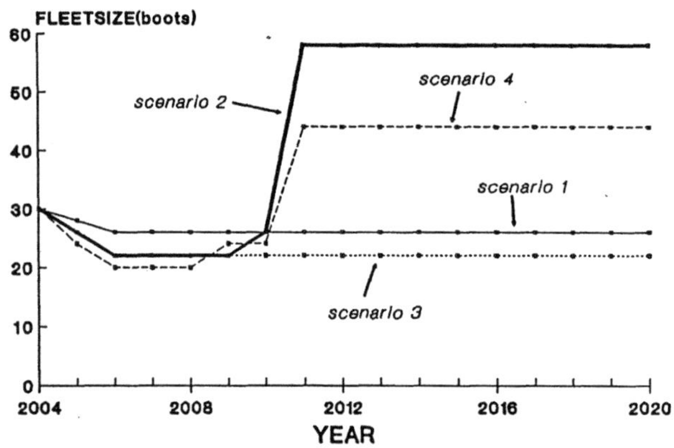


**Fig. 14.20.** Fleet size development of NGFS

## 14.7 Conclusions

The current Aegean maritime network is strongly regulated by the Greek government. Only Greek businesses are allowed to operate within the network, in accordance with a licence system. Moreover, there is a certain degree of subsidization of routes that are not commercially viable, but that must be served for reasons of accessibility of all inhabited Aegean islands (equity motive). The conventional ferries are still largely serving the market. However, on a number of routes hydrofoils are increasingly being used.

The technological innovations in the shipping sector in the last decade have resulted in increased speed, lower pollution levels and more luxurious and more silent ships. This higher attractivity will have a positive impact on demand for maritime transport. The faster the research into and the technological development of new ships goes, the sooner the traditional ferries will be out of business. The pace of introduction of more comfortable, larger and faster ships should not be underestimated.

The completion of the internal market of the European Union (EU) due to the European unification process implies that from 2004 onwards, non-Greek EU shipping companies will also be allowed to exploit routes in the Aegean. The cabotage privilege of Greek companies will then be lifted. Competition on profitable lines will increase significantly. Accordingly, the price per passenger kilometre will drop and the quality of service is likely to improve. Shipping companies will try to do their utmost to maintain or even to increase their market share. Therefore, they have the possibility to differentiate themselves from their competitors, by price and service. Greek companies will have to show a more commercial attitude in serving the (free) market; otherwise non-Greek companies will benefit from the lack of the use of marketing techniques by Greek companies. More attention for passenger needs will guarantee or even strengthen market shares of the companies at hand. The level of competition will be Damocles' sword above the heads of the weakest companies. Although there will be an unbalanced market in the year after 2004, we predict a stabilization of market shares in the long run.

If the political relations of Greece with Turkey improve, economic relations are also likely to ameliorate. Additionally, numerous forms of travel restrictions will be relieved. This will lead to an increase in Greek-Turkish traffic, favouring the transport sector. This will also shift the spatial distribution of demand for transport in the Aegean. The centre of gravity of the network will probably gradually move eastward and a more centrally situated hub island will be more viable and yield significant network efficiency gains.

One of the possibilities in the hands of the network actors and operators to improve the attractivity of the network is to design a hub-and-spokes type of network with Paros as a potential hub. This type of network bears a larger degree of connection density and accordingly a shorter average travel time. This is especially beneficial to the transport users due to the supply of a more attractive service. The transport line has to face a loss in the production of passenger kilometres, but the generated efficiency gains will increase demand.

The model sketched above appeared to yield a plausible description of the Aegean maritime transport market. It also has a sizeable set of extension possibilities. We will discuss here only a few. A first model improvement can be found in the direction of statistical analysis by investigating the probability of appearance of various scenarios. Second, the improvement of the relationships between the variables may be realized by undertaking a more comprehensive causal analysis. Third, it is interesting to consider also air transport, the service quality, the structure of the transport lines and more price differentiation and market segmentation. Finally, the model would become more realistic by using proper econometric methods and by undertaking data analysis to identify more proper values of the parameters. In the present version of the model we have used the method of calibration and 'questination' because of missing data in order to get plausible parameters.

## References


Biehl D (1986) Infrastructure and Regional Development. EC, Brussels

Bruinsma FR (1994) De Invloed van Transportinfrastructuur. Nederlandse Geografische Studies. 175 Amsterdam

Capello R (1994) Telecommunication and Regional Development. Avebury, Aldershot, UK

Cuadrado-Roura JR, Nijkamp P, Salva P (Eds) (1994) Moving Frontiers. Avebury, Aldershot, UK

ECMT (1993) Short Sea Shipping. Paris

Goulielmos AM, Lekakos M (1992) The economic structure of Greek passenger coastal shipping. In: Proceedings of the First European Research Roundtable Conference on Shortsea Shipping. Technical University Delft, Lloyd's of London Press Ltd. 301-315

Gwilliam KM (Ed) (1993) Current Issues Maritime Economics. Kluwer, Dordrecht

Hiemstra F, Zwier R (1994) Passenger Shipping System in the Aegean, Master's Thesis. Dept. of Economics and Econometrics, Free University, Amsterdam

Jacobsen U, Kristiansen J (1992) Transport networks for peripheral regions. In: Transportation research, 198-210

Lloyd's List International (1991) Greece and Champion of the cabotage. London, 8-46

Nijkamp P (Ed) (1994) New Borders and Old Barriers in Spatial Development. Avebury, Aldershot, UK

Nijkamp P, Reichman S, Wegener S (Eds) (1990) Euromobile. Gower, Aldershot, UK

Nijkamp P, Vleugel J (1994) Missing Transport Networks in Europe. Avebury, Aldershot, UK

Porter ME (1990) The Competitive Advantage of Nations. Free Press, Boston,

Psaraftis HN (1993) Prospects for the conventional passenger/car ferry in the Aegean. Paper presented at Conference, Overcoming isolation. Plomari, Lesbos, Greece

Psaraftis HN, Papanikolaou AD (1992) Impact of new technologies on shortsea shipping in Greece. National Technical University of Athens. Paper presented at First European Research Roundtable Conference on Shortsea Shipping. TU Delft, The Netherlands

Wijnolst N, Peeters C, Liebman P (Eds) (1993) European Short Sea Shipping. Lloyd's of London Press

# 15 Transport Networks and Insular Isolation: Measuring Spatial Inequality

S. Kostopoulou

## 15.1 Introduction

Insular regions, whether they are large islands or complexes of small islands, being strictly bounded regions have geographical characteristics that differentiate them from mainland regions: land discontinuity, distance from the mainland and sparse location at sea. These characteristics constitute factors of spatial inequality, which increases in relation to the insular problems.

Regarding the transport sector, geographical implications result in extremely difficult conditions, affecting the social and economic development of the islands in such a way that they are rendered disadvantageous areas and compose an essential state of environment: isolation.

The meaning of isolation is highly problematic to define and much more to measure. In spatial analysis the term 'isolation' is employed in order to define specific situations and geographical areas. The state of spatial isolation is usually regarded as a disadvantageous one, characterising remote and problematic areas. Regional planning policies aim to reduce and overcome isolation and therefore provide a basis for the development of remote areas.

In the planning decision process, measuring the degree of isolation by formulating appropriate indexes, could be a useful tool. Evaluating the level of isolation of each region in relation to central and regional transport networks, could lead to the identification of areas with special needs in transport infrastructure and services. Thus, the degree of isolation could be considered as a measure of spatial inequality.

This paper is an attempt to interpret the nature and content of this concept and to propose its measurement.

## 15.2 The Concept of Isolation

The basic meaning of isolation is separation: it is the action and the result of separating, preventing communication with others. Isolation may be interpreted in different ways, according to whether it is considered as a conclusive or descriptive term, definitive concept, conscious feeling or part of the social structure. The term is used in all social sciences, at different levels of analysis and in relation to different concepts. (Gould and Kolb 1972)

In sociology, isolation is defined as 'the cutting off or minimization of communication between groups or societies', based on religion, class and other criteria. In demography, a population unit is considered to be in a state of isolation when all its members are forced to choose their spouses from members of the group. In political science the term isolation or isolationism implies a national policy based on the assumption that, it is better for the interest of the state that all international relations are eliminated. In economic science, the term 'isolationism' refers to the national economic policy of a closed economy. In this sense, isolation implies a policy of economic autarchy by minimizing the use of imported goods.

In the geographical context, the term 'isolation' characterizes a distinctive land surface separated from others by natural barriers. In the broader spatial context, it implies 'distance from' and more generally, difficulty of communication with other places.

However, isolation is not simply a function of spatial separation, related to communication facilities; it is influenced by many factors, which act simultaneously and interact at several levels. Stone (1972) provided a list of factors classified as follows:

a. Physical elements
- Geographical bounds or barriers (islands, natural vegetation barriers, 'deep' in woods, coastal configuration, gorge barriers) which form ubiquitous constraints and impose additional costs to the construction of infrastucture networks.
- Climate (cut off by snow, floods, cancellation of scheduled transport services due to bad weather conditions).
- Resources (lack of natural fuels, lack of those natural resources that would help form self-sufficient communities)

b. Social elements
- Facilities (insufficient health, education services)
- Transport (low level or lack of transport facilities)
- Communication (insufficient communication links and information)

c. Cultural elements
- Minorities (due to language, race, religion, nationality, political beliefs)
- Economic situation (extreme cases of poverty)
- Physical handicaps, socially marginal groups
- Psychological factors

The contribution of these factors to the interpretation of the state of isolation, depends closely upon the scale of phenomena under consideration. At the local level, the importance of psychological and cultural elements and their impact on people, is emphasised; at the regional level geographical implications (physical, social) and their impact on spatial units (communities, regions), are considered as vital factors. (Nutley 1980)

When one examines isolation at a regional level, the main reference is to the possibilities of communication between regions. From this point of view, isolation is related to the concept of accessibility, which is generally defined as the possibility of using the opportunities that the economic, social, cultural and political facilities and institutions provide. (Domanski 1979) Isolation can thus be regarded as the lowest level of a continuous scale of accessibility, which refers to a low level or even complete lack of communication.

The ease of access varies among people, social groups or places. Differentiation is due to many factors (income, level of culture), among them the geographical. Considering only the latter, we refer to spatial accessibility in relation to distance and transport networks.

Spatial accessibility contains two basic dimensions: the possibility to reach a place by a transport mode (the connection component) and the cost of being reached quickly and/or cheaply (the movement cost component). The accessibility of one place to another depends upon two factors: whether or not they are connected by a transport route and the magnitude of the movement cost, which is usually related to distance. (Cox 1972)

For the measurement of accessibility a large number of indicators have been developed; in general, they seek to measure the distance between human activities or communities that are connected with one another by means of a transport system. The main components of any accessibility indicator are: the measurement of the distance between any two points, the connection with the communication system which makes it possible to cover the distance between them, the effort, in time or cost, required to cover the distance, and the connection with the particular type of activity in which the user wishes to take part. Accessibility indicators are used in the planning and evaluation of transport, urban and regional development plans, in relation to other socio-economic or regional factors. (Morris, Dumble and Wigal 1979; Pirie 1979)

Measuring the level of accessibility of each area may then facilitate the identification of their differences and contribute to the determination of spatial inequalities. Inequalities in spatial accessibility of different regions relate to inequalities in transport networks, which are due to the geographical peculiarities of each region (mountainous, insular), as well as to the existing transport infrastructure, resulting from approved transport policies.

In Europe, the main aim of the transport policy during the first decades after the Second World War, was the development and improvement of existing communications infrastucture, without anticipating for new, more homogeneous and a balanced regional structure of transport networks. (Izquierdo 1992) As a result, a cumulative process was introduced that resulted in a greater concentration of traffic on the improved networks, which in its turn, required further extensions. This process contributed to larger regional imbalances, and gave rise to the centre-periphery pattern.

Areas belonging to the periphery are usually located far from urban centers and have a low level of accessibility to communication networks. Peripheral regions display a low level of quality of life, a less developed economic structures, a small proportion of capital-technology intensive industries and phenomena of socio-cultural and politico-administrative dependence. (Commission of the European Communities 1981)

The most problematic peripheral regions are the most remote ones with an extremely low level of accessibility, that is those that have to deal with the problem of isolation. Insular regions could be considered as a particular, or even extreme case of peripheral region, and insular isolation could be defined as a disadvantageous situation of spatial inequality.

## 15.3 Insular Isolation

Insular isolation is characterized by acute transport network problems, due to the special geographic configuration of the islands. These peculiar spatial conditions constitute negative factors for the development of the islands. Specifically they:

- restrict the possibilities of movement to the mainland and the island centers because of the dependence upon public transport only.

- affect positively the cost and negatively the quality of transport services because of the problems caused by the structure of transport networks, i.e. number of intermediate terminals, frequency irregularities of scheduled services.

- increase the production cost of insular products, thus limiting their distribution only to regional and local markets, a fact that prevents the creation of economies of scale because of the small market size.

- raise the cost of living and form a negative psychological environment for insular societies, that affects directly or indirectly all human activities and, therefore, the quality of life of the inhabitants.

The intensity of these provlems is proportional to the size of islands: large islands have better opportunities for development, as well as means of political pressure on the centres of decision making, for their benefit. Conversely, small islands, because of their limited population and the size of their markets, do not dispose the comparative advantages, which would promote their development. (Communautes Europeennes 1988)

In the transport sector, small and sparsely populated island complexes, seriously increase the relative construction costs of appropriate infrastructure. The provision of satisfactory transport services is much more expensive per user. These restrictions usually cause complications and transport problems in small islands: dramatically higher transport cost, insufficiencies in/or lack of infrastructure, lack of adequate transport connections, extended journey time, frequency irregularities of scheduled services.

These problems reduce the opportunities for economic activities and furthermore limit the ease of access. Thus, they cause migration, especially affecting the young population, shrinking the development capacity of the islands even more and reinforcing the state of isolation.

## 15.4 Measure of Insular Isolation

The state of isolation of an insular region may be analyzed at three levels: local, regional, inter-regional.

- *Local level.* Isolation is estimated within the island, that is the level of accessibility of the inhabitants of each community (town, village) as a spatial unit, to the island's airport and/or port. In the case of large islands, movement is dependent on the quality of road and/or rail transport networks, the frequency of public transport services and car ownership. In the case of very small islands, without internal transport, ease of access to the port depends upon the size of the island and the available transport means. (Nutley 1980)
- *Regional level.* Isolation is estimated separately for every island as a spatial unit, in relation to the possibility of getting to the mainland and other islands. The level of accessibility depends upon the distance from the mainland and insular transport centres and the quality of transport services, i.e. existing infrastucture in ports and airports, number of sea and air connections, frequency of scheduled services.
- *Inter-regional level.* Isolation is assessed for the insular region as a whole, in relation to other mainland regions within a national or broader context. The level of accessibility here depends upon the region's connectivity to national and international transport networks, infrastructure capacity, frequency and quality of services, in comparison with those of the mainland regions.

## 15.5 Insular Isolation Coefficient

We attempt here to measure insular isolation at a regional level. For this purpose, an Insular Isolation Coefficient is suggested, for the measurement of each island's accessibility to the mainland and insular transport nodes (ports, airports).

Following the aforementioned analysis, the Insular Isolation Coefficient is considered to depend upon several factors, which may be grouped into the following categories:

A. Geographical
   Distance from the nearest mainland/insular ports/airports
B. Topological
   Existence/level of infrastructure in ports/airports
C. Transportation
   Number of sea/air links between mainland/insular ports/airports
D. Economic
   Costs of sea/air trips to the nearest mainland/insular ports/airports
E. Qualitative
   Time-distances of sea/air trips to the nearest mainland/insular ports/airports.

F. Psychological
  a. Number of sea / air connections cancelled per year
  b. Maximum time of lack of all transport connections per year.

A weight is then given to each factor, which expresses the relative importance and impact of the factor to the island's accessibility level.

After evaluating the isolation coefficient of each island, we may then proceed to their clustering in categories, and draw a map of accessibility areas. This clustering will lead to the detection of island groups with different transport needs and could therefore become a useful tool for the design of insular transport networks.

In this first approach, the three first categories are considered in the proposed Insular Isolation Coefficient (IIC), which will be examined to the air and sea transport network (IICA and IICS respectively). Specifically, the items that define the factors for each transport mode are as follows:

*Air Transport Network*

A. Geographical factors
  1. Distance from the nearest airport
B. Topological factors
  2. Level of infrastructure: small, medium, large airport
C. Transportation factors
  3. Number of airports directly connected with the island.
  3a. Number of mainland airports directly connected with the island
  3b. Number of insular airports directly connected with the island
  4. Number of air connections per week
  4a. Number of air connections with mainland airports per week
  4b. Number of air connections with insular airports per week.

*Sea Transport Network*

A. Geographical factors
  1. Distance from the nearest mainland port.
B. Topological factors
  2. Level of infrastructure: small, medium, large port.
C. Transportation factors
  3. Number of ports connected with the island
  3a. Number of mainland ports connected with the island
  3b. Number of insular ports connected with the island
  4. Number of sea connections per week
  4a. Number of sea connections with mainland ports per week
  4b. Number of sea connections with insular ports per week
  5. Number of direct sea connections per week with mainland and/or insular ports

The above can be expressed in graph theory terms as follows: (Carre 1979)

*Air Transport Network Graph*

A = (Q, U) : weighted connected graph (insular air transport network)

$Q = \{q_1, q_2, \ldots\ldots, q_r+_l\}$ where $q_i$: insular or mainland airport

$Q = X \cup Z$ where $X = \{x_1,\ldots,x_r\}$, $Z = \{z_1,\ldots\ldots,z_l\}$ symbolizes the distinct insular and mainland airports respectively

$U = \{(x_i,q_j)\}$ : set of edges $(x_i,q_j)$ expressing air transport links

$k(x_i,q_j$ : weight of edge $(x_i,q_j)$ expressing the number of connections between incident nodes per week

$dt(x_i)$ : distance of node $x_i$ from the closest mainland or insular airport

$w_1$ : degree of importance of item $dt(x_i)$, $w_1$ E {1,2,3}, where 1 means low, 2 means medium, 3 means high.

$df(x_i)$ : infrastructure degree of node $x_i$ (airport capacity)

$w_2$ : degree of importance of item $df(x_i)$, $w_2$ E {1,2,3}

$d(x_i)$ : number of incident nodes to $x_i$

$w_3$ : degree of importance of item $d(x_i)$, $w_3$ E {1,2,3}

$dr(x_i)$ : number of regional incident nodes to $x_i$
where: regional nodes = mainland airports

$w_4$ : degree of importance of item $dr(x_i)$, $w_4$ E {1,2,3}

$dl(x_i)$ : number of local incident nodes to $x_i$
where: local nodes = insular airports
Obviously $d(x_i) = dr(x_i) + dl(x_i)$

$w_5$ : degree of importance of item $dl(x_i)$, $w_5$ E {1,2,3}

$dc(x_i)$ : connectivity degree of node $x_i$
(sum of weights of incident edges)

$w_6$ : degree of importance of item $dc(x_i)$, $w_6$ E {1,2,3}

$drc(x_i)$ : regional connectivity degree of node $x_i$
(sum of weights of regional incident edges
where: regional edges = connections with mainland airports)

$w_7$ : degree of importance of item $drc(x_i)$, $w_7$ E {1,2,3}

$dlc(x_i)$ : local connectivity degree of node $x_i$
(sum of weights of local incident edges
where: local edges = connections with insular airports)

$w_8$ : degree of importance of item $dlc(x_i)$, $w_8$ E {1,2,3}

$IICA(x_i)$ : Insular Isolation Coefficient of node $x_i$ in the air transport network

$$IICA(x_i) = w_1\, dt(x_i) + w_2\, df(x_i) + w_3\, d(x_i) + w_4\, dr(x_i) + w_5\, dl(x_i) + w_6\, dc(x_i) + w_7\, drc(x_i) + w_8\, dlc(x_i)$$

*Sea Transport Network Graph*

S = (Q, U) : weighted connected graph (insular sea transport network)

$Q = \{q_1, q_2, \ldots\ldots, q_r+_l\}$ where $q_i$ : insular or mainland port

$Q = X \cup Z$ where $X = \{x_1,\ldots,x_r\}$, $Z = \{z_1,\ldots\ldots,z_l\}$ symbolizes the distinct insular and mainland ports respectively

$U = \{(x_i,q_j)\}$: set of edges $(x_i,q_j)$ expressing sea transport links

$k(x_i,q_j)$ : weight of edge $(x_i,q_j)$ expressing the number of sea connections between incident nodes per week

$dt(x_i)$ : distance of node $x_i$ from the closest mainland port

| | |
|---|---|
| $w_1$ | : degree of importance of item $dt(x_i)$, $w_1$ E {1,2,3}, where 1 means low, 2 means medium, 3 means high. |
| $df(x_i)$ | : infrastructure degree of node $x_i$ (port capacity) |
| $w_2$ | : degree of importance of item $df(x_i)$, $w_2$ E {1,2,3} |
| $d(x_i)$ | : number of incident nodes to $x_i$ |
| $w_3$ | : degree of importance of item $d(x_i)$, $w_3$ E {1,2,3} |
| $dr(x_i)$ | : number of regional paths through node $x_i$ where: regional paths = mainland starting paths |
| $w_4$ | : degree of importance of item $dr(x_i)$, $w_4$ E {1,2,3} |
| $dl(x_i)$ | : number of local paths through node $x_i$ where: local paths = insular starting paths |
| $w_5$ | : degree of importance of item $dl(x_i)$, $w_5$ E {1,2,3} |
| $dc(x_i)$ | : connectivity degree of node $x_i$ (sum of weights of incident edges) |
| $w_6$ | : degree of importance of item $dc(x_i)$, $w_6$ E {1,2,3} |
| $drc(x_i)$ | : regional connectivity degree of node $x_i$ (sum of weights of incident edges of regional paths) |
| $w_7$ | : degree of importance of item $drc(x_i)$, $w_7$ E {1,2,3} |
| $dlc(x_i)$ | : local connectivity degree of node $x_i$ (sum of weights of incident edges of local paths) |
| $w_8$ | : degree of importance of item $dlc(x_i)$, $w_8$ E {1,2,3} |
| $IICS(x_i)$ | : Insular Isolation Coefficient of node $x_i$ in the sea transport network |
| $IICS(x_i)$ | $= w_1\, dt(x_i) + w_2\, df(x_i) + w_3\, d(x_i) + w_4\, dr(x_i) + w_5\, dl(x_i) + w_6\, dc(x_i) + w_7\, drc(x_i) + w_8\, dlc(x_i)$ |

## 15.6 The Study Area

The Insular Isolation Coefficient method is the Aegean archipel, which is the largest insular region of Greece in size and population. The Aegean region is characterized by its peripheral location, large number of islands, sparse population, communication difficulties and spatial variations in economic development. We examine all inhabited islands of the five prefectures of the Aegean Region (prefectures of Lesvos, Chios, Samos, Dodecanese and Cyclades), the Sporades Islands (administratively belonging to the prefecture of Magnesia) and the island of Samothrace (administratively belonging to the prefecture of Evros). (Table 15.1, Map 15.1)

The air and sea transport netorks are presented in Figures 15.1 and 15.2 show the air and sea transport networks respectively. Tables 15.2 and 15.3 show the values of the items that define the geographical, topological and transportation factors of the Air and Sea Insular Isolation Coefficient respectively. Transportation data are based on Olympic Airways and Olympic Aviation winter 1993 timetables, and on sea ferries February 1993 timetables.

Table 15.1.

| $x_i$ | Island name | Surface ($Km^2$) | Population (1981) |
|---|---|---|---|
| $x_1$ | Agathonisi | 13,5 | 133 |
| $x_2$ | Ag.Efstratios | 43,0 | 296 |
| $x_3$ | Alonnisos | 65,0 | 1.528 |
| $x_4$ | Amorgos | 121,0 | 1.718 |
| $x_5$ | Anafi | 38,0 | 292 |
| $x_6$ | Andros | 401,0 | 5.020 |
| $x_7$ | Antiparos | 35,0 | 635 |
| $x_8$ | Arki | 6,7 | 68 |
| $x_9$ | Astipalea | 97,0 | 1.030 |
| $x_{10}$ | Dilos | 3,4 | 16 |
| $x_{11}$ | Donousa | 14,0 | 116 |
| $x_{12}$ | Iraklia | 13,0 | 95 |
| $x_{13}$ | Thira | 76,0 | 7.083 |
| $x_{14}$ | Thiresia | 9,3 | 245 |
| $x_{15}$ | Thimena | 9,9 | 136 |
| $x_{16}$ | Ikaria | 255,0 | 7.559 |
| $x_{17}$ | Ios | 108,0 | 1.451 |
| $x_{18}$ | Kalimnos | 111,0 | 14.295 |
| $x_{19}$ | Karpathos | 301,0 | 4.645 |
| $x_{20}$ | Kasos | 66,0 | 1.134 |
| $x_{21}$ | Kea | 131,0 | 1.648 |
| $x_{22}$ | Kimolos | 36,0 | 786 |
| $x_{23}$ | Koufonisia | 6,0 | 232 |
| $x_{24}$ | Kithnos | 99,0 | 1.502 |
| $x_{25}$ | Kos | 290,0 | 20.350 |
| $x_{26}$ | Lipsi | 26,0 | 574 |
| $x_{27}$ | Leros | 53,0 | 8.127 |
| $x_{28}$ | Lesvos | 1.630,0 | 88.601 |
| $x_{29}$ | Limnos | 476,0 | 15.721 |
| $x_{30}$ | Megisti | 9,0 | 222 |
| $x_{31}$ | Milos | 151,0 | 4.554 |
| $x_{32}$ | Mikonos | 86,0 | 5.503 |
| $x_{33}$ | Naxos | 429,0 | 14.037 |
| $x_{34}$ | Nisiros | 41,0 | 916 |
| $x_{35}$ | Inousses | 14,2 | 703 |
| $x_{36}$ | Paros | 193,0 | 7.881 |
| $x_{37}$ | Patmos | 34,0 | 2.534 |
| $x_{38}$ | Rinia | 13,7 | 11 |
| $x_{39}$ | Rhodes | 1.401,0 | 87.831 |
| $x_{40}$ | Samothrace | 173,0 | 2.871 |
| $x_{41}$ | Samos | 476,0 | 31.629 |
| $x_{42}$ | Serifos | 73,2 | 1.133 |
| $x_{43}$ | Sikinos | 41,0 | 290 |
| $x_{44}$ | Sifnos | 73,0 | 2.087 |
| $x_{45}$ | Skiathos | 48,0 | 4.127 |
| $x_{46}$ | Skopelos | 96,0 | 4.451 |
| $x_{47}$ | Skiros | 209,5 | 2.757 |
| $x_{48}$ | Symi | 58,0 | 2.273 |
| $x_{49}$ | Syros | 34,0 | 19.668 |
| $x_{50}$ | Schinoussa | 7,8 | 140 |
| $x_{51}$ | Tilos | 62,8 | 300 |
| $x_{52}$ | Tinos | 195,0 | 7.730 |
| $x_{53}$ | Folegandros | 32,0 | 547 |
| $x_{54}$ | Fourni | 30,0 | 1.203 |
| $x_{55}$ | Chalki | 28,0 | 334 |
| $x_{56}$ | Chios | 840,0 | 49.865 |
| $x_{57}$ | Psara | 39,8 | 460 |
| $x_{58}$ | Pserimos | 14,8 | 72 |

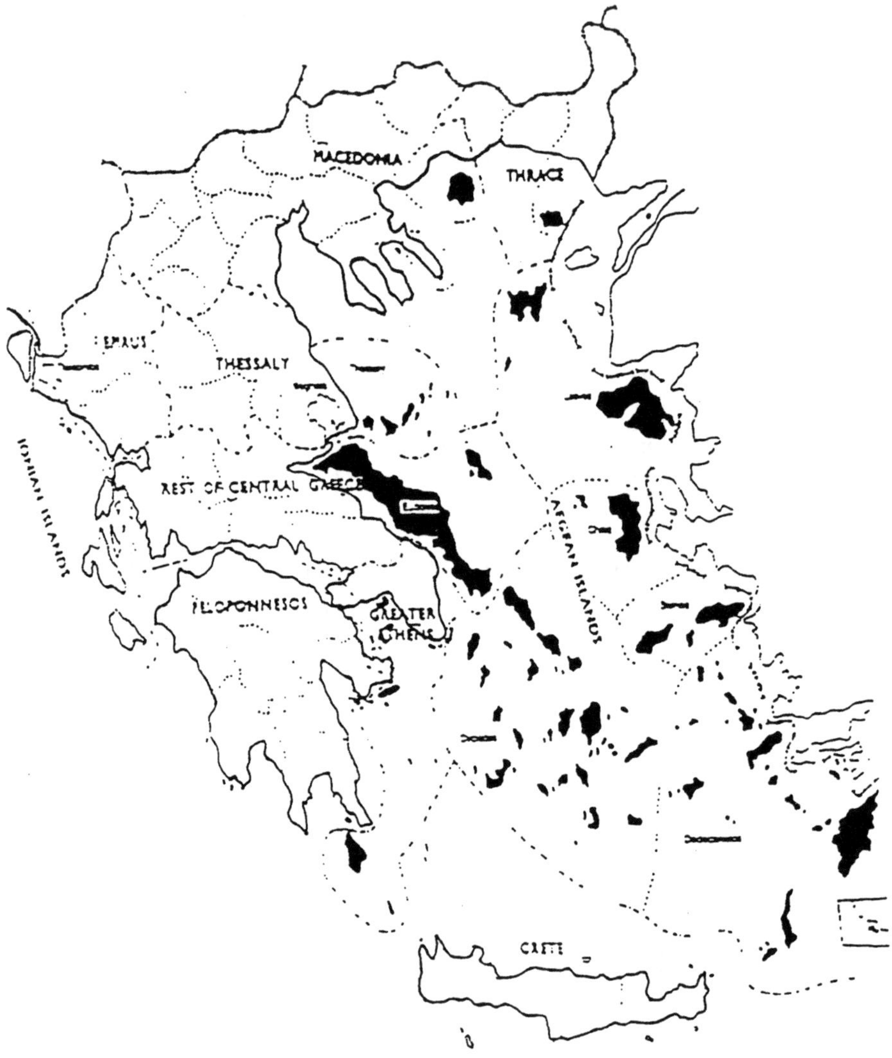


**Map 15.1.** Aegean Islands

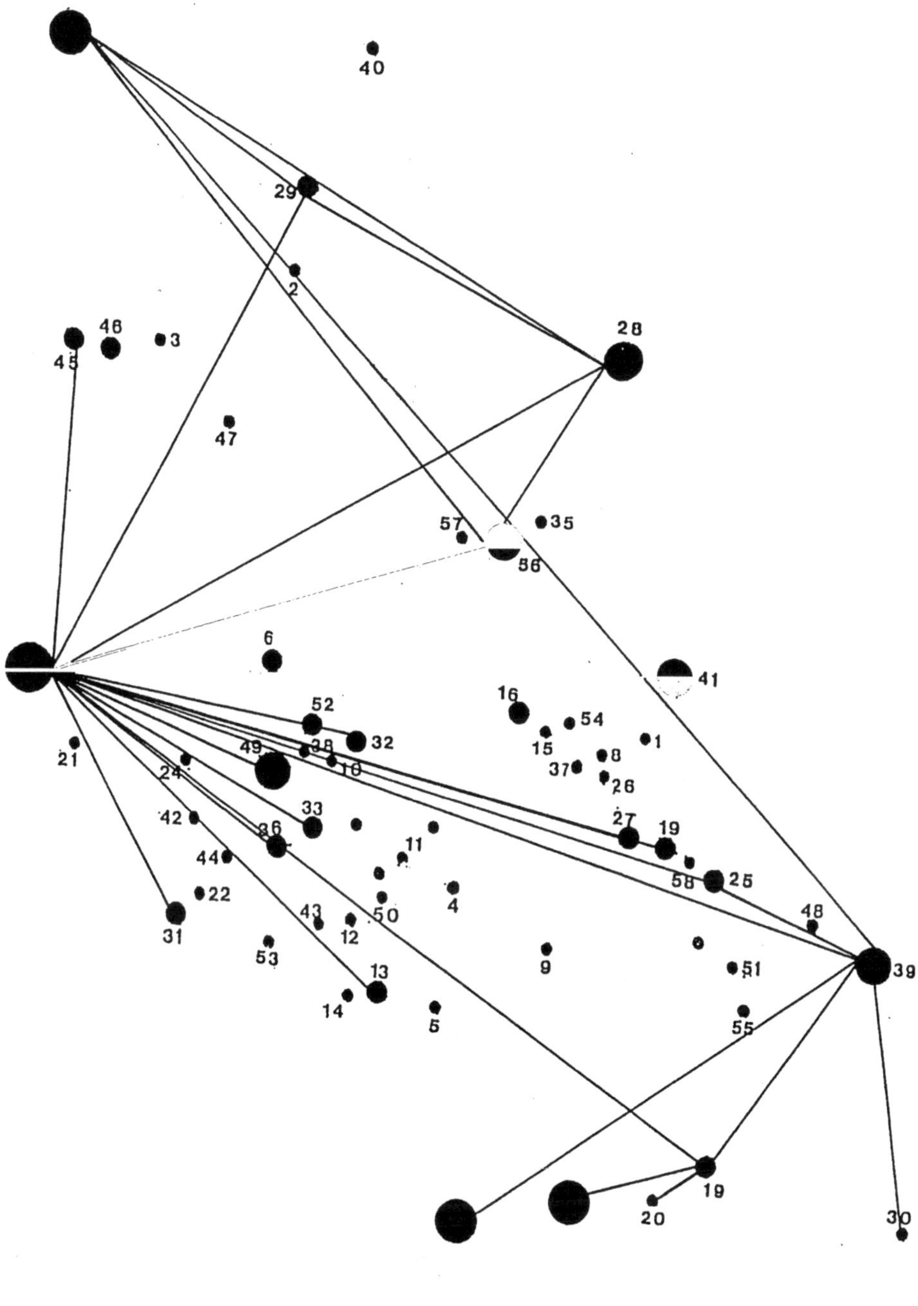


Fig. 15.1. Air transport network

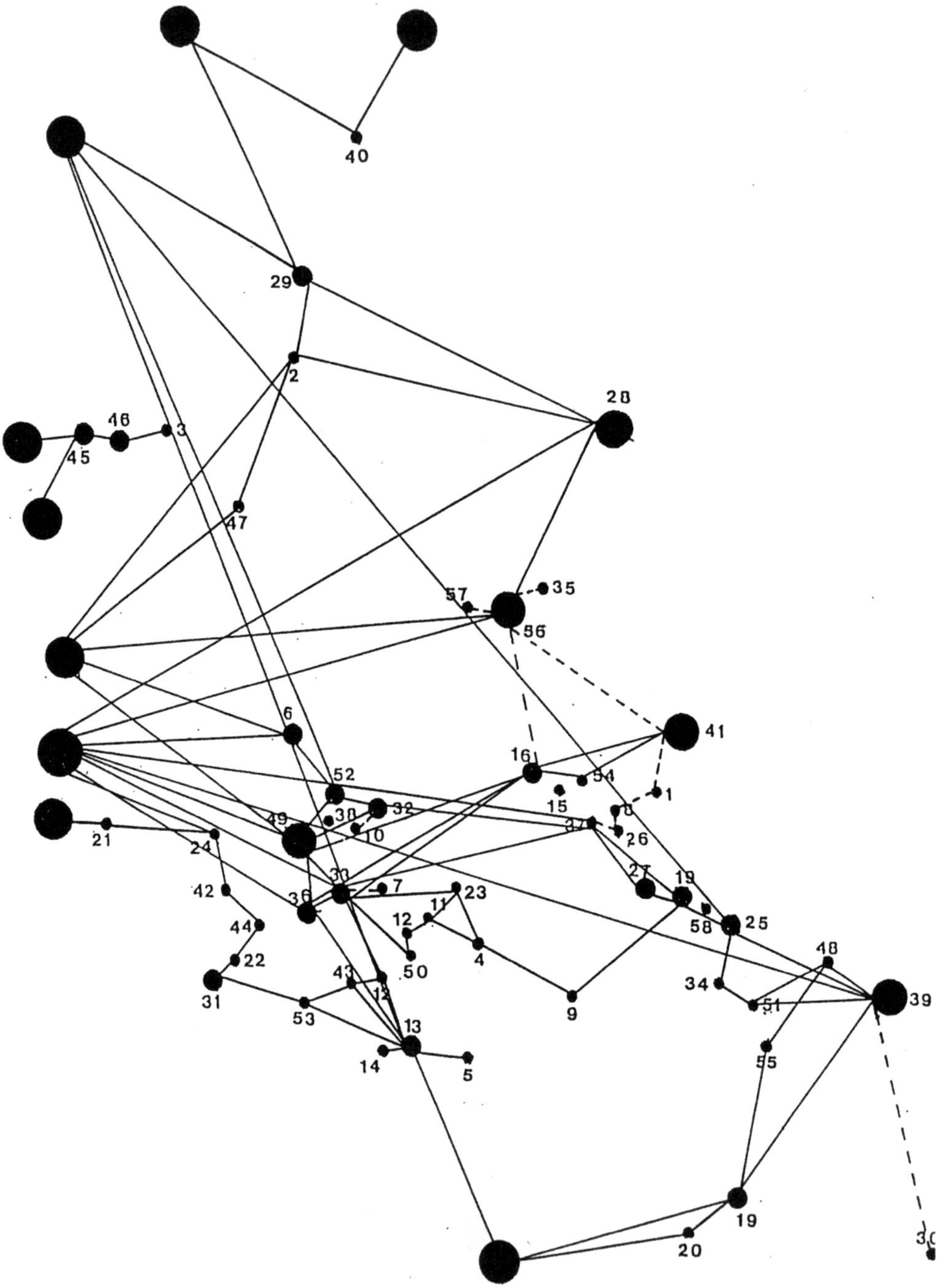


Fig. 15.2. Sea transport network

**Table 15.2.** Values of items air transport network

| $x_i$ | Island name | $dt_i$ | $df_i$ | $d_i$ | $dr_i$ | $dl_i$ | $dc_i$ | $drc_i$ | $dlc_i$ |
|---|---|---|---|---|---|---|---|---|---|
| $x_1$ | Agathonisi | 25 | 0 | 0 | 0 | 0 | 0 | 0 | 0 |
| $x_2$ | Ag.Efstratios | 35 | 0 | 0 | 0 | 0 | 0 | 0 | 0 |
| $x_3$ | Alonnisos | 25 | 0 | 0 | 0 | 0 | 0 | 0 | 0 |
| $x_4$ | Amorgos | 40 | 0 | 0 | 0 | 0 | 0 | 0 | 0 |
| $x_5$ | Anafi | 25 | 0 | 0 | 0 | 0 | 0 | 0 | 0 |
| $x_6$ | Andros | 35 | 0 | 0 | 0 | 0 | 0 | 0 | 0 |
| $x_7$ | Antiparos | 5 | 0 | 0 | 0 | 0 | 0 | 0 | 0 |
| $x_8$ | Arki | 25 | 0 | 0 | 0 | 0 | 0 | 0 | 0 |
| $x_9$ | Astipalea | 0 | 1 | 0 | 0 | 0 | 0 | 0 | 0 |
| $x_{10}$ | Dilos | 5 | 0 | 0 | 0 | 0 | 0 | 0 | 0 |
| $x_{11}$ | Donousa | 30 | 0 | 0 | 0 | 0 | 0 | 0 | 0 |
| $x_{12}$ | Iraklia | 25 | 0 | 0 | 0 | 0 | 0 | 0 | 0 |
| $x_{13}$ | Thira | 0 | 3 | 1 | 1 | 0 | 41 | 41 | 0 |
| $x_{14}$ | Thiresia | 5 | 0 | 0 | 0 | 0 | 0 | 0 | 0 |
| $x_{15}$ | Thimena | 40 | 0 | 0 | 0 | 0 | 0 | 0 | 0 |
| $x_{16}$ | Ikaria | 50 | 0 | 0 | 0 | 0 | 0 | 0 | 0 |
| $x_{17}$ | Ios | 30 | 0 | 0 | 0 | 0 | 0 | 0 | 0 |
| $x_{18}$ | Kalimnos | 20 | 0 | 0 | 0 | 0 | 0 | 0 | 0 |
| $x_{19}$ | Karpathos | 0 | 2 | 4 | 1 | 3 | 16 | 2 | 14 |
| $x_{20}$ | Kasos | 0 | 1 | 1 | 0 | 1 | 3 | 0 | 3 |
| $x_{21}$ | Kea | 40 | 0 | 0 | 0 | 0 | 0 | 0 | 0 |
| $x_{22}$ | Kimolos | 10 | 0 | 0 | 0 | 0 | 0 | 0 | 0 |
| $x_{23}$ | Koufonisia | 30 | 0 | 0 | 0 | 0 | 0 | 0 | 0 |
| $x_{24}$ | Kithnos | 40 | 0 | 0 | 0 | 0 | 0 | 0 | 0 |
| $x_{25}$ | Kos | 0 | 3 | 2 | 1 | 1 | 14 | 12 | 2 |
| $x_{26}$ | Lipsi | 15 | 0 | 0 | 0 | 0 | 0 | 0 | 0 |
| $x_{27}$ | Leros | 0 | 1 | 1 | 1 | 0 | 4 | 4 | 0 |
| $x_{28}$ | Lesvos | 0 | 3 | 4 | 2 | 2 | 33 | 27 | 6 |
| $x_{29}$ | Limnos | 0 | 3 | 3 | 2 | 1 | 22 | 18 | 4 |
| $x_{30}$ | Megisti | 0 | 1 | 1 | 0 | 1 | 2 | 0 | 2 |
| $x_{31}$ | Milos | 0 | 1 | 1 | 1 | 0 | 7 | 7 | 0 |
| $x_{32}$ | Mikonos | 0 | 2 | 1 | 1 | 0 | 11 | 11 | 0 |
| $x_{33}$ | Naxos | 0 | 1 | 1 | 1 | 0 | 14 | 14 | 0 |
| $x_{34}$ | Nisiros | 30 | 0 | 0 | 0 | 0 | 0 | 0 | 0 |
| $x_{35}$ | Inousses | 15 | 0 | 0 | 0 | 0 | 0 | 0 | 0 |
| $x_{36}$ | Paros | 0 | 1 | 1 | 1 | 0 | 14 | 14 | 0 |
| $x_{37}$ | Patmos | 25 | 0 | 0 | 0 | 0 | 0 | 0 | 0 |
| $x_{38}$ | Rinia | 5 | 0 | 0 | 0 | 0 | 0 | 0 | 0 |
| $x_{39}$ | Rhodes | 0 | 3 | 6 | 2 | 4 | 48 | 32 | 16 |
| $x_{40}$ | Samothrace | 40 | 0 | 0 | 0 | 0 | 0 | 0 | 0 |
| $x_{41}$ | Samos | 0 | 3 | 1 | 1 | 0 | 14 | 14 | 0 |
| $x_{42}$ | Serifos | 35 | 0 | 0 | 0 | 0 | 0 | 0 | 0 |
| $x_{43}$ | Sikinos | 35 | 0 | 0 | 0 | 0 | 0 | 0 | 0 |
| $x_{44}$ | Sifnos | 30 | 0 | 0 | 0 | 0 | 0 | 0 | 0 |
| $x_{45}$ | Skiathos | 0 | 2 | 1 | 1 | 0 | 3 | 3 | 0 |
| $x_{46}$ | Skopelos | 20 | 0 | 0 | 0 | 0 | 0 | 0 | 0 |
| $x_{47}$ | Skiros | 0 | 2 | 0 | 0 | 0 | 0 | 0 | 0 |
| $x_{48}$ | Symi | 15 | 0 | 0 | 0 | 0 | 0 | 0 | 0 |
| $x_{49}$ | Syros | 0 | 1 | 1 | 1 | 0 | 14 | 14 | 0 |
| $x_{50}$ | Schinoussa | 25 | 0 | 0 | 0 | 0 | 0 | 0 | 0 |
| $x_{51}$ | Tilos | 35 | 0 | 0 | 0 | 0 | 0 | 0 | 0 |
| $x_{52}$ | Tinos | 10 | 0 | 0 | 0 | 0 | 0 | 0 | 0 |
| $x_{53}$ | Folegandros | 30 | 0 | 0 | 0 | 0 | 0 | 0 | 0 |
| $x_{54}$ | Fourni | 40 | 0 | 0 | 0 | 0 | 0 | 0 | 0 |
| $x_{55}$ | Chalki | 50 | 0 | 0 | 0 | 0 | 0 | 0 | 0 |
| $x_{56}$ | Chios | 0 | 2 | 3 | 2 | 1 | 22 | 20 | 2 |
| $x_{57}$ | Psara | 50 | 0 | 0 | 0 | 0 | 0 | 0 | 0 |
| $x_{58}$ | Pserimos | 10 | 0 | 0 | 0 | 0 | 0 | 0 | 0 |

**Table 15.3.** Values of items sea transport network

| $x_i$ | Island name | $dt_i$ | $df_i$ | $d_i$ | $dr_i$ | $dl_i$ | $dc_i$ | $drc_i$ | $dlc_i$ |
|---|---|---|---|---|---|---|---|---|---|
| $x_1$ | Agathonisi | 177 | 1 | 2 | 0 | 6 | 2 | 0 | 2 |
| $x_2$ | Ag.Efstratios | 51 | 2 | 4 | 2 | 3 | 3 | 3 | 0 |
| $x_3$ | Alonnisos | 27 | 2 | 1 | 2 | 2 | 40 | 40 | 0 |
| $x_4$ | Amorgos | 136 | 2 | 3 | 1 | 14 | 5 | 3 | 2 |
| $x_5$ | Anafi | 139 | 1 | 1 | 1 | 5 | 2 | 2 | 0 |
| $x_6$ | Andros | 89 | 3 | 2 | 1 | 2 | 7 | 7 | 0 |
| $x_7$ | Antiparos | 75 | 1 | 1 | 0 | 1 | 14 | 0 | 14 |
| $x_8$ | Arki | 157 | 1 | 2 | 0 | 7 | 2 | 0 | 2 |
| $x_9$ | Astipalea | 169 | 2 | 2 | 1 | 13 | 3 | 2 | 1 |
| $x_{10}$ | Dilos | 93 | 0 | 1 | 0 | 1 | 1 | 0 | 1 |
| $x_{11}$ | Donousa | 110 | 2 | 3 | 1 | 14 | 5 | 3 | 2 |
| $x_{12}$ | Iraklia | 102 | 1 | 2 | 1 | 8 | 3 | 1 | 2 |
| $x_{13}$ | Thira | 130 | 3 | 7 | 2 | 21 | 27 | 19 | 8 |
| $x_{14}$ | Thiresia | 118 | 1 | 1 | 0 | 1 | 2 | 0 | 2 |
| $x_{15}$ | Thimena | 140 | 0 | 0 | 0 | 0 | 0 | 0 | 0 |
| $x_{16}$ | Ikaria | 143 | 3 | 5 | 1 | 7 | 9 | 7 | 2 |
| $x_{17}$ | Ios | 111 | 3 | 3 | 1 | 8 | 21 | 13 | 8 |
| $x_{18}$ | Kalimnos | 183 | 3 | 4 | 1 | 19 | 30 | 8 | 27 |
| $x_{19}$ | Karpathos | 242 | 3 | 4 | 1 | 14 | 3 | 3 | 0 |
| $x_{20}$ | Kasos | 238 | 2 | 2 | 1 | 14 | 2 | 2 | 0 |
| $x_{21}$ | Kea | 45 | 2 | 2 | 1 | 1 | 7 | 7 | 0 |
| $x_{22}$ | Kimolos | 86 | 2 | 2 | 1 | 13 | 3 | 3 | 0 |
| $x_{23}$ | Koufonisia | 106 | 1 | 2 | 1 | 14 | 2 | 2 | 2 |
| $x_{24}$ | Kithnos | 52 | 2 | 3 | 2 | 15 | 4 | 4 | 0 |
| $x_{25}$ | Kos | 200 | 3 | 4 | 2 | 18 | 28 | 7 | 21 |
| $x_{26}$ | Lipsi | 158 | 1 | 2 | 0 | 6 | 2 | 0 | 2 |
| $x_{27}$ | Leros | 171 | 3 | 2 | 1 | 8 | 6 | 4 | 2 |
| $x_{28}$ | Lesvos | 188 | 3 | 4 | 4 | 3 | 5 | 5 | 0 |
| $x_{29}$ | Limnos | 84 | 3 | 4 | 4 | 3 | 5 | 5 | 0 |
| $x_{30}$ | Megisti | 320 | 2 | 1 | 0 | 2 | 2 | 0 | 2 |
| $x_{31}$ | Milos | 87 | 2 | 2 | 1 | 13 | 3 | 3 | 0 |
| $x_{32}$ | Mikonos | 94 | 3 | 5 | 3 | 8 | 31 | 22 | 9 |
| $x_{33}$ | Naxos | 103 | 3 | 7 | 1 | 27 | 29 | 21 | 8 |
| $x_{34}$ | Nisiros | 197 | 2 | 2 | 1 | 12 | 4 | 2 | 2 |
| $x_{35}$ | Inousses | 126 | 1 | 1 | 0 | 1 | 6 | 0 | 6 |
| $x_{36}$ | Paros | 97 | 3 | 7 | 2 | 30 | 34 | 26 | 8 |
| $x_{37}$ | Patmos | 163 | 2 | 5 | 1 | 15 | 9 | 7 | 2 |
| $x_{38}$ | Rinia | 70 | 0 | 1 | 0 | 0 | 1 | 0 | 1 |
| $x_{39}$ | Rhodes | 250 | 3 | 6 | 2 | 28 | 15 | 13 | 2 |
| $x_{40}$ | Samothrace | 29 | 3 | 1 | 1 | 0 | 6 | 6 | 0 |
| $x_{41}$ | Samos | 175 | 3 | 4 | 1 | 13 | 11 | 7 | 4 |
| $x_{42}$ | Serifos | 73 | 3 | 2 | 1 | 16 | 4 | 3 | 1 |
| $x_{43}$ | Sikinos | 95 | 1 | 3 | 1 | 19 | 4 | 3 | 1 |
| $x_{44}$ | Sifnos | 79 | 3 | 2 | 1 | 16 | 4 | 3 | 1 |
| $x_{45}$ | Skiathos | 41 | 3 | 3 | 2 | 2 | 41 | 41 | 0 |
| $x_{46}$ | Skopelos | 43 | 3 | 2 | 2 | 2 | 41 | 41 | 0 |
| $x_{47}$ | Skiros | 53 | 3 | 3 | 2 | 2 | 8 | 8 | 0 |
| $x_{48}$ | Symi | 230 | 2 | 3 | 1 | 23 | 2 | 2 | 0 |
| $x_{49}$ | Syros | 83 | 3 | 7 | 2 | 27 | 25 | 24 | 1 |
| $x_{50}$ | Schinoussa | 53 | 1 | 2 | 1 | 8 | 3 | 1 | 2 |
| $x_{51}$ | Tilos | 212 | 2 | 3 | 1 | 14 | 2 | 2 | 0 |
| $x_{52}$ | Tinos | 86 | 3 | 2 | 1 | 5 | 15 | 15 | 0 |
| $x_{53}$ | Folegandros | 87 | 1 | 3 | 1 | 13 | 4 | 3 | 1 |
| $x_{54}$ | Fourni | 141 | 2 | 2 | 1 | 6 | 4 | 2 | 2 |
| $x_{55}$ | Chalki | 232 | 2 | 2 | 1 | 13 | 1 | 1 | 0 |
| $x_{56}$ | Chios | 110 | 3 | 5 | 4 | 7 | 8 | 4 | 6 |
| $x_{57}$ | Psara | 70 | 2 | 1 | 0 | 1 | 2 | 0 | 2 |
| $x_{58}$ | Pserimos | 187 | 0 | 0 | 0 | 0 | 0 | 0 | 0 |

**Table 15.4.** Classification of the islands (air transport network)

| Rank | $x_i$ | Island name | $w_1$ $dt_i$ | $w_2$ $df_i$ | $w_3$ $d_i$ | $w_4$ $dr_i$ | $w_5$ $dl_i$ | $w_6$ $dc_i$ | $w_7$ $drc_i$ | $w_8$ $dlc_i$ | $IICA_i$ |
|---|---|---|---|---|---|---|---|---|---|---|---|
| 1 | $x_{16}$ | Ikaria | 1/3 | 0 | 0 | 0 | 0 | 0 | 0 | 0 | 1/3 |
| 2 | $x_{55}$ | Chalki | 1/3 | 0 | 0 | 0 | 0 | 0 | 0 | 0 | 1/3 |
| 3 | $x_{57}$ | Psara | 1/3 | 0 | 0 | 0 | 0 | 0 | 0 | 0 | 1/3 |
| 4 | $x_4$ | Amorgos | 1/3 | 0 | 0 | 0 | 0 | 0 | 0 | 0 | 1/3 |
| 5 | $x_{15}$ | Thymena | 1/3 | 0 | 0 | 0 | 0 | 0 | 0 | 0 | 1/3 |
| 6 | $x_{21}$ | Kea | 1/3 | 0 | 0 | 0 | 0 | 0 | 0 | 0 | 1/3 |
| 7 | $x_{24}$ | Kythnos | 1/3 | 0 | 0 | 0 | 0 | 0 | 0 | 0 | 1/3 |
| 8 | $x_{40}$ | Samothrace | 1/3 | 0 | 0 | 0 | 0 | 0 | 0 | 0 | 1/3 |
| 9 | $x_{54}$ | Fourni | 1/3 | 0 | 0 | 0 | 0 | 0 | 0 | 0 | 1/3 |
| 10 | $x_2$ | Ag.Efstratios | 1/3 | 0 | 0 | 0 | 0 | 0 | 0 | 0 | 1/3 |
| 11 | $x_6$ | Andros | 1/3 | 0 | 0 | 0 | 0 | 0 | 0 | 0 | 1/3 |
| 12 | $x_{42}$ | Serifos | 1/3 | 0 | 0 | 0 | 0 | 0 | 0 | 0 | 1/3 |
| 13 | $x_{43}$ | Sikinos | 1/3 | 0 | 0 | 0 | 0 | 0 | 0 | 0 | 1/3 |
| 14 | $x_{51}$ | Tilos | 1/3 | 0 | 0 | 0 | 0 | 0 | 0 | 0 | 1/3 |
| 15 | $x_{17}$ | Ios | 1/2 | 0 | 0 | 0 | 0 | 0 | 0 | 0 | 1/2 |
| 16 | $x_{23}$ | Koufonisia | 1/2 | 0 | 0 | 0 | 0 | 0 | 0 | 0 | 1/2 |
| 17 | $x_{34}$ | Nisiros | 1/2 | 0 | 0 | 0 | 0 | 0 | 0 | 0 | 1/2 |
| 18 | $x_{11}$ | Donoussa | 1/2 | 0 | 0 | 0 | 0 | 0 | 0 | 0 | 1/2 |
| 19 | $x_{44}$ | Sifnos | 1/2 | 0 | 0 | 0 | 0 | 0 | 0 | 0 | 1/2 |
| 20 | $x_{53}$ | Folegandros | 1/2 | 0 | 0 | 0 | 0 | 0 | 0 | 0 | 1/2 |
| 21 | $x_1$ | Agathonisi | 1/2 | 0 | 0 | 0 | 0 | 0 | 0 | 0 | 1/2 |
| 22 | $x_3$ | Alonnisos | 1/2 | 0 | 0 | 0 | 0 | 0 | 0 | 0 | 1/2 |
| 23 | $x_5$ | Anafi | 1/2 | 0 | 0 | 0 | 0 | 0 | 0 | 0 | 1/2 |
| 24 | $x_8$ | Arki | 1/2 | 0 | 0 | 0 | 0 | 0 | 0 | 0 | 1/2 |
| 25 | $x_{12}$ | Hiraklia | 1/2 | 0 | 0 | 0 | 0 | 0 | 0 | 0 | 1/2 |
| 26 | $x_{37}$ | Patmos | 1/2 | 0 | 0 | 0 | 0 | 0 | 0 | 0 | 1/2 |
| 27 | $x_{50}$ | Schinoussa | 1/2 | 0 | 0 | 0 | 0 | 0 | 0 | 0 | 1/2 |
| 28 | $x_{18}$ | Kalimnos | 1 | 0 | 0 | 0 | 0 | 0 | 0 | 0 | 1 |
| 29 | $x_{46}$ | Skopelos | 1 | 0 | 0 | 0 | 0 | 0 | 0 | 0 | 1 |
| 30 | $x_{26}$ | Lipsi | 1 | 0 | 0 | 0 | 0 | 0 | 0 | 0 | 1 |
| 31 | $x_{35}$ | Inousses | 1 | 0 | 0 | 0 | 0 | 0 | 0 | 0 | 1 |
| 32 | $x_{48}$ | Symi | 1 | 0 | 0 | 0 | 0 | 0 | 0 | 0 | 1 |
| 33 | $x_{22}$ | Kimolos | 1 | 0 | 0 | 0 | 0 | 0 | 0 | 0 | 1 |
| 34 | $x_{52}$ | Tinos | 1 | 0 | 0 | 0 | 0 | 0 | 0 | 0 | 1 |
| 35 | $x_{58}$ | Pserimos | 1 | 0 | 0 | 0 | 0 | 0 | 0 | 0 | 1 |
| 36 | $x_7$ | Antiparos | 1 | 0 | 0 | 0 | 0 | 0 | 0 | 0 | 1 |
| 37 | $x_{10}$ | Dilos | 1 | 0 | 0 | 0 | 0 | 0 | 0 | 0 | 1 |
| 38 | $x_{14}$ | Thiresia | 1 | 0 | 0 | 0 | 0 | 0 | 0 | 0 | 1 |
| 39 | $x_{38}$ | Rinia | 1 | 0 | 0 | 0 | 0 | 0 | 0 | 0 | 1 |
| 40 | $x_9$ | Astypalea | 1 | 3 | 0 | 0 | 0 | 0 | 0 | 0 | 4 |
| 41 | $x_{47}$ | Skyros | 1 | 6 | 0 | 0 | 0 | 0 | 0 | 0 | 7 |
| 42 | $x_{30}$ | Megisti | 1 | 3 | 1 | 0 | 2 | 2 | 0 | 4 | 13 |
| 43 | $x_{20}$ | Kasos | 1 | 3 | 1 | 0 | 2 | 3 | 0 | 6 | 16 |
| 44 | $x_{45}$ | Skiathos | 1 | 6 | 1 | 3 | 0 | 3 | 9 | 0 | 23 |
| 45 | $x_{27}$ | Leros | 1 | 3 | 1 | 3 | 0 | 4 | 12 | 0 | 24 |
| 46 | $x_{31}$ | Milos | 1 | 3 | 1 | 3 | 0 | 7 | 21 | 0 | 36 |
| 47 | $x_{32}$ | Mykonos | 1 | 6 | 1 | 3 | 0 | 11 | 33 | 0 | 55 |
| 48 | $x_{33}$ | Naxos | 1 | 3 | 1 | 3 | 0 | 14 | 42 | 0 | 64 |
| 49 | $x_{36}$ | Paros | 1 | 3 | 1 | 3 | 0 | 14 | 42 | 0 | 64 |
| 50 | $x_{49}$ | Syros | 1 | 3 | 1 | 3 | 0 | 14 | 42 | 0 | 64 |
| 51 | $x_{19}$ | Karpathos | 1 | 6 | 4 | 3 | 6 | 16 | 6 | 28 | 70 |
| 52 | $x_{41}$ | Samoς | 1 | 9 | 1 | 3 | 0 | 14 | 42 | 0 | 70 |
| 53 | $x_{25}$ | Kos | 1 | 9 | 2 | 3 | 2 | 14 | 36 | 4 | 71 |
| 54 | $x_{56}$ | Chios | 1 | 6 | 3 | 6 | 2 | 22 | 60 | 4 | 104 |
| 55 | $x_{29}$ | Limnos | 1 | 9 | 3 | 6 | 2 | 22 | 54 | 8 | 105 |
| 56 | $x_{28}$ | Lesvos | 1 | 9 | 4 | 6 | 4 | 33 | 81 | 12 | 150 |
| 57 | $x_{13}$ | Thira | 1 | 9 | 1 | 3 | 0 | 41 | 123 | 0 | 178 |
| 58 | $x_{39}$ | Rhodes | 1 | 9 | 6 | 6 | 8 | 48 | 96 | 32 | 206 |

**Table 15.5.** Classification of the islands (sea transport network)

| Rank | $x_i$ | Island name | $w_1$ $dt_i$ | $w_2$ $df_i$ | $w_3$ $d_i$ | $w_3$ $dr_i$ | $w_5$ $dl_i$ | $w_6$ $dc_i$ | $w_7$ $drc_i$ | $w_8$ $dlc_i$ | $IICS_i$ |
|---|---|---|---|---|---|---|---|---|---|---|---|
| 1 | $x_{58}$ | Pserimos | 1/2 | 0 | 0 | 0 | 0 | 0 | 0 | 0 | 1/2 |
| 2 | $x_{15}$ | Thimena | 1/2 | 0 | 0 | 0 | 0 | 0 | 0 | 0 | 1/2 |
| 3 | $x_{38}$ | Rinia | 1 | 0 | 1 | 0 | 0 | 1 | 0 | 1 | 4 |
| 4 | $x_{10}$ | Dilos | 1 | 0 | 1 | 0 | 1 | 1 | 0 | 1 | 5 |
| 5 | $x_{14}$ | Thiresia | 1/2 | 3 | 1 | 0 | 1 | 2 | 0 | 2 | 9 1/2 |
| 6 | $x_{57}$ | Psara | 1 | 6 | 1 | 0 | 1 | 2 | 0 | 2 | 13 |
| 7 | $x_{30}$ | Megisti | 1/3 | 6 | 1 | 0 | 2 | 2 | 0 | 2 | 13 1/3 |
| 8 | $x_1$ | Agathonisi | 1/2 | 3 | 2 | 0 | 6 | 2 | 0 | 2 | 15 1/2 |
| 9 | $x_{26}$ | Lipsi | 1/2 | 3 | 2 | 0 | 6 | 2 | 0 | 2 | 15 1/2 |
| 10 | $x_8$ | Arki | 1/2 | 3 | 2 | 0 | 7 | 2 | 0 | 2 | 16 1/2 |
| 11 | $x_{35}$ | Inousses | 1/2 | 3 | 1 | 0 | 1 | 6 | 0 | 6 | 17 1/2 |
| 12 | $x_5$ | Anafi | 1/2 | 3 | 1 | 3 | 5 | 2 | 6 | 0 | 20 1/2 |
| 13 | $x_{12}$ | Iraklia | 1/2 | 3 | 2 | 3 | 8 | 3 | 3 | 2 | 24 1/2 |
| 14 | $x_{50}$ | Schinoussa | 1 | 3 | 2 | 3 | 8 | 3 | 3 | 2 | 25 |
| 15 | $x_{55}$ | Chalki | 1/3 | 6 | 2 | 3 | 13 | 1 | 3 | 0 | 28 1/3 |
| 16 | $x_{54}$ | Fourni | 1/2 | 6 | 2 | 3 | 6 | 4 | 6 | 2 | 29 1/2 |
| 17 | $x_2$ | Ag.Efstratios | 1 | 6 | 4 | 6 | 3 | 3 | 9 | 0 | 32 |
| 18 | $x_{23}$ | Koufonisia | 1/2 | 3 | 2 | 3 | 14 | 2 | 6 | 2 | 32 1/2 |
| 19 | $x_{20}$ | Kasos | 1/3 | 6 | 2 | 3 | 14 | 2 | 6 | 0 | 33 1/3 |
| 20 | $x_7$ | Antiparos | 1 | 3 | 1 | 0 | 1 | 14 | 0 | 14 | 34 |
| 21 | $x_{51}$ | Tilos | 1/3 | 6 | 3 | 3 | 14 | 2 | 6 | 0 | 34 1/3 |
| 22 | $x_9$ | Astipalea | 1/2 | 6 | 2 | 3 | 13 | 3 | 6 | 1 | 34 1/2 |
| 23 | $x_{34}$ | Nisiros | 1/2 | 6 | 2 | 3 | 12 | 4 | 6 | 2 | 35 1/2 |
| 24 | $x_{22}$ | Kimolos | 1 | 6 | 2 | 3 | 13 | 3 | 9 | 0 | 37 |
| 25 | $x_{31}$ | Milos | 1 | 6 | 2 | 3 | 13 | 3 | 9 | 0 | 37 |
| 26 | $x_{53}$ | Folegandros | 1 | 3 | 3 | 3 | 13 | 4 | 9 | 1 | 37 |
| 27 | $x_{40}$ | Samothrace | 1 | 9 | 1 | 3 | 0 | 6 | 18 | 0 | 38 |
| 28 | $x_{21}$ | Kea | 1 | 6 | 2 | 3 | 1 | 7 | 21 | 0 | 41 |
| 29 | $x_{19}$ | Karpathos | 1/3 | 9 | 4 | 3 | 14 | 3 | 9 | 0 | 42 1/3 |
| 30 | $x_4$ | Amorgos | 1/2 | 6 | 3 | 3 | 14 | 5 | 9 | 2 | 42 1/2 |
| 31 | $x_{11}$ | Donousa | 1/2 | 6 | 3 | 3 | 14 | 5 | 9 | 2 | 42 1/2 |
| 32 | $x_{27}$ | Leros | 1/2 | 9 | 2 | 3 | 8 | 6 | 12 | 2 | 42 1/2 |
| 33 | $x_{43}$ | Sikinos | 1 | 3 | 3 | 3 | 19 | 4 | 9 | 1 | 43 |
| 34 | $x_{48}$ | Symi | 1/3 | 6 | 3 | 3 | 23 | 2 | 6 | 0 | 43 1/3 |
| 35 | $x_{42}$ | Serifos | 1 | 9 | 2 | 3 | 16 | 4 | 9 | 1 | 45 |
| 36 | $x_{44}$ | Sifnos | 1 | 9 | 2 | 3 | 16 | 4 | 9 | 1 | 45 |
| 37 | $x_6$ | Andros | 1 | 9 | 2 | 3 | 2 | 7 | 21 | 0 | 45 |
| 38 | $x_{24}$ | Kithnos | 1 | 6 | 3 | 6 | 15 | 4 | 12 | 0 | 47 |
| 39 | $x_{28}$ | Lesvos | 1/2 | 9 | 4 | 12 | 3 | 5 | 15 | 0 | 48 1/2 |
| 40 | $x_{29}$ | Limnos | 1 | 9 | 4 | 12 | 3 | 5 | 15 | 0 | 49 |
| 41 | $x_{47}$ | Skiros | 1 | 9 | 3 | 6 | 2 | 8 | 24 | 0 | 53 |
| 42 | $x_{16}$ | Ikaria | 1/2 | 9 | 5 | 3 | 7 | 9 | 21 | 2 | 56 1/2 |
| 43 | $x_{56}$ | Chios | 1/2 | 9 | 5 | 12 | 7 | 8 | 12 | 6 | 59 1/2 |
| 44 | $x_{37}$ | Patmos | 1/2 | 6 | 5 | 3 | 15 | 9 | 21 | 2 | 61 1/2 |
| 45 | $x_{41}$ | Samos | 1/2 | 9 | 4 | 3 | 13 | 11 | 21 | 4 | 65 1/2 |
| 46 | $x_{52}$ | Tinos | 1 | 9 | 2 | 3 | 5 | 15 | 45 | 0 | 80 |
| 47 | $x_{17}$ | Ios | 1/2 | 9 | 3 | 3 | 8 | 21 | 39 | 8 | 91 1/2 |
| 48 | $x_{39}$ | Rhodes | 1/3 | 9 | 6 | 6 | 28 | 15 | 39 | 2 | 105 1/3 |
| 49 | $x_{25}$ | Kos | 1/3 | 9 | 4 | 6 | 18 | 28 | 21 | 21 | 107 1/3 |
| 50 | $x_{18}$ | Kalimnos | 1/2 | 9 | 4 | 3 | 19 | 30 | 21 | 27 | 113 1/2 |
| 51 | $x_{13}$ | Thira | 1/2 | 9 | 7 | 6 | 21 | 27 | 57 | 8 | 127 1/2 |
| 52 | $x_{32}$ | Mikonos | 1 | 9 | 5 | 9 | 8 | 31 | 66 | 9 | 13[illegible] |
| 53 | $x_{33}$ | Naxos | 1/2 | 9 | 7 | 3 | 27 | 29 | 63 | 8 | 146 1/2 |
| 54 | $x_{49}$ | Syros | 1 | 9 | 7 | 6 | 27 | 25 | 72 | 1 | 14[illegible] |
| 55 | $x_{36}$ | Paros | 1 | 9 | 7 | 6 | 30 | 34 | 78 | 8 | 17[illegible] |
| 56 | $x_3$ | Alonnisos | 1 | 6 | 1 | 6 | 2 | 40 | 120 | 0 | 17[illegible] |
| 57 | $x_{46}$ | Skopelos | 1 | 9 | 2 | 6 | 2 | 41 | 123 | 0 | 18[illegible] |
| 58 | $x_{45}$ | Skiathos | 1 | 9 | 3 | 6 | 2 | 41 | 123 | 0 | 18[illegible] |

Table 15.6.

| Rank | $x_i$ | Island name | Population (1981) | Surface ($Km^2$) | IICA | IICS |
|---|---|---|---|---|---|---|
| 1 | $x_{38}$ | Rinia | 11 | 13,7 | 1 | 4 |
| 2 | $x_{10}$ | Dilos | 16 | 3,4 | 1 | 5 |
| 3 | $x_{8}$ | Arki | 68 | 6,7 | 1/2 | 16 1/2 |
| 4 | $x_{58}$ | Pserimos | 72 | 14,8 | 1 | 1/2 |
| 5 | $x_{12}$ | Iraklia | 95 | 13,0 | 1/2 | 24 1/2 |
| 6 | $x_{11}$ | Donousa | 116 | 14,0 | 1/2 | 42 1/2 |
| 7 | $x_{1}$ | Agathonisi | 133 | 13,5 | 1/2 | 15 1/2 |
| 8 | $x_{15}$ | Thimena | 136 | 9,9 | 1/3 | 1/2 |
| 9 | $x_{50}$ | Schinoussa | 140 | 7,8 | 1/2 | 25 |
| 10 | $x_{30}$ | Megisti | 222 | 9,0 | 13 | 13 1/3 |
| 11 | $x_{23}$ | Koufonisia | 232 | 6,0 | 1/2 | 32 1/2 |
| 12 | $x_{14}$ | Thiresia | 245 | 9,3 | 1 | 9 1/2 |
| 13 | $x_{43}$ | Sikinos | 290 | 41,0 | 1/3 | 43 |
| 14 | $x_{5}$ | Anafi | 292 | 38,0 | 1/2 | 20 1/2 |
| 15 | $x_{2}$ | Ag.Efstratios | 296 | 43,0 | 1/3 | 32 |
| 16 | $x_{51}$ | Tilos | 300 | 62,8 | 1/3 | 34 1/3 |
| 17 | $x_{55}$ | Chalki | 334 | 28,0 | 1/3 | 28 1/3 |
| 18 | $x_{57}$ | Psara | 460 | 39,8 | 1/3 | 13 |
| 19 | $x_{53}$ | Folegandros | 547 | 32,0 | 1/2 | 37 |
| 20 | $x_{26}$ | Lipsi | 574 | 26,0 | 1 | 15 1/2 |
| 21 | $x_{7}$ | Antiparos | 635 | 35,0 | 1 | 34 |
| 22 | $x_{35}$ | Inousses | 703 | 14,2 | 1 | 17 1/2 |
| 23 | $x_{22}$ | Kimolos | 786 | 36,0 | 1 | 37 |
| 24 | $x_{34}$ | Nisiros | 916 | 41,0 | 1/2 | 35 1/2 |
| 25 | $x_{9}$ | Astipalea | 1.030 | 97,0 | 4 | 34 1/2 |
| 26 | $x_{42}$ | Serifos | 1.133 | 73,2 | 1/3 | 45 |
| 27 | $x_{20}$ | Kasos | 1.134 | 66,0 | 16 | 33 1/3 |
| 28 | $x_{54}$ | Fourni | 1.203 | 30,0 | 1/3 | 29 1/2 |
| 29 | $x_{17}$ | Ios | 1.451 | 108,0 | 1/2 | 91 1/2 |
| 30 | $x_{24}$ | Kithnos | 1.502 | 99,0 | 1/3 | 47 |
| 31 | $x_{3}$ | Alonnisos | 1.528 | 65,0 | 1/2 | 176 |
| 32 | $x_{21}$ | Kea | 1.648 | 131,0 | 1/3 | 41 |
| 33 | $x_{4}$ | Amorgos | 1.718 | 121,0 | 1/3 | 42 1/2 |
| 34 | $x_{44}$ | Sifnos | 2.087 | 73,0 | 1/2 | 45 |
| 35 | $x_{48}$ | Symi | 2.273 | 58,0 | 1 | 43 1/3 |
| 36 | $x_{37}$ | Patmos | 2.534 | 34,0 | 1/2 | 61 1/2 |
| 37 | $x_{47}$ | Skiros | 2.757 | 209,5 | 7 | 53 |
| 38 | $x_{40}$ | Samothrace | 2.871 | 173,0 | 1/3 | 38 |
| 39 | $x_{45}$ | Skiathos | 4.127 | 48,0 | 23 | 185 |
| 40 | $x_{46}$ | Skopelos | 4.451 | 96,0 | 1 | 184 |
| 41 | $x_{31}$ | Milos | 4.554 | 151,0 | 36 | 37 |
| 42 | $x_{19}$ | Karpathos | 4.645 | 301,0 | 70 | 42 1/3 |
| 43 | $x_{6}$ | Andros | 5.020 | 401,0 | 1/3 | 45 |
| 44 | $x_{32}$ | Mikonos | 5.503 | 86,0 | 55 | 138 |
| 45 | $x_{13}$ | Thira | 7.083 | 76,0 | 178 | 127 |
| 46 | $x_{16}$ | Ikaria | 7.559 | 255,0 | 1/3 | 56 1/2 |
| 47 | $x_{52}$ | Tinos | 7.730 | 195,0 | 1 | 80 |
| 48 | $x_{36}$ | Paros | 7.881 | 193,0 | 64 | 173 |
| 49 | $x_{27}$ | Leros | 8.127 | 53,0 | 24 | 42 1/2 |
| 50 | $x_{33}$ | Naxos | 14.037 | 429,0 | 64 | 146 1/2 |
| 51 | $x_{18}$ | Kalimnos | 14.295 | 111,0 | 1 | 113 1/2 |
| 52 | $x_{29}$ | Limnos | 15.721 | 476,0 | 105 | 49 |
| 53 | $x_{49}$ | Syros | 19.668 | 84,0 | 64 | 149 |
| 54 | $x_{25}$ | Kos | 20.350 | 290,0 | 71 | 107 1/3 |
| 55 | $x_{41}$ | Samos | 31.629 | 476,0 | 70 | 65 1/2 |
| 56 | $x_{56}$ | Chios | 49.865 | 840,0 | 104 | 59 1/2 |
| 57 | $x_{39}$ | Rhodes | 87.831 | 1.401,0 | 206 | 105 1/3 |
| 58 | $x_{28}$ | Lesvos | 88.601 | 1.630,0 | 150 | 48 1/2 |

Next, weights are then given to each factor, according to their relative importance to the island's accessibility, as follows :

Air transport network
$w_1$ : 0 - 24 = 1, 25 - 35 = 1/2 36 - 50 = 1/3
$w_2 = 3$, $w_3 = 1$, $w_4 = 3$, $w_5 = 2$, $w_6 = 1$, $w_7 = 3$, $w_8 = 2$

Sea transport network
$w_1$ : 1 - 100 = 1, 101 - 200 = 1/2, 201 - 320 = 1/3
$w_2 = 3$, $w_3 = 1$, $w_4 = 3$, $w_5 = 1$, $w_6 - 1$, $w_7 = 3$, $w_8 = 1$

Tables 15.4 and 15.5 show the classification of the islands according to the Insular Isolation Coefficients IICA and IICS.

## 15.7 Conclusions

Our results show that in the Aegean region there are islands with a highly satisfactory, a medium and an extremely low level of accessibility.

By reference to Table 15.1 (island's surface and population), we can see that in general, large islands belong to the first two groups, and small islands to the last (Table 15.6). Small, sparsely populated islands are usually suffering from isolation. They do not have equal opportunities of access to mainland transport networks. Furthermore, they do not have equal opportunities in labour, health services, education, leisure, information. These communities experience isolation as a phenomenon of spatial inequality, spatial injustice.

Any attempt to overcome spatial inequality by providing the necessary transport infrastructure to all islands, imposes high costs to small, sparse populations. Economic efficiency as a criterion in transport network design, here is opposite to spatial equity in transport services, causing a planning decision dilemma: economic transport planning or social transport planning.

Stressing our interest in the latter we must remember that the right to access is an essential and invaluable human right in a Democracy. The state has to provide access equally to all citizens. Furthermore, spatial separation must be alleviated by providing all necessary transport services. Therefore, the attempt to overcome isolation in disadvantageous insular regions becomes the main aim of social regional policy and the main criterion in insular transport network design.

## References

Carre B (1979) Graphs and networks. Clarendon Press, Oxford
Commission of the European Communities (1981) Europe 2000. Outlook for the development of the Community's territory

Communautes Européennes (1988) Regions insulaires defavorisées, Comité Economique et Social
Cox RK (1972) Man, Location and Behavior. An Introduction to Human Geography. John Wiley and Sons, New York
Domanski R (1979) Accessibility, efficiency, and spatial organization. Environment and Planning A 11:1189-1206
Gould J, Kolb W (Eds) (1972) Dictionary of Social Sciences. UNESCO
Izquierdo R (1992) Infrastructure capacity and network access. 12th International Symposium of the European Conference of Ministers of Transport on Theory and Practice in Transport Economics. Lisbon, May 1992
Nutley SD (1980) The concept of 'isolation'. A method of evaluation and a West Highland Example. Regional Studies 14:111-123
Morris JM, Dumble PL, Wigan MR (1979) Accessibility indicators for transport planning. Transportation Research A 13:91-109
Pirie GH (1979) Measuring accessibility. A review and proposal, Environment and Planning 11:299-312
Stone KH (1972) The concepts of isolation, Geoforum 11:74 -77

# 16 Overcoming Isolation and the Role of Transport: The Case of the Aegean Islands

G. A. Giannopoulos and G. Boulougaris

## 16.1 Introduction

There are a large number of areas within the Community and the various other European countries, that suffer from poor accessibility with regard to central locations. This condition is invariably related to poor economic performance in the same regions, depopulation, etc. A term which has come to be used synonymously with all these conditions is 'peripheral' regions (or even countries). The notion of 'peripherality' is directly related to being on the 'periphery' with regard to a central area or centre, i.e. on the outskirts or circumference (as indeed in the original meaning of the Greek word 'peripheria' - periphery). In the context of regional development issues, periphery and peripheral have come to be today, almost synonymous with depressed, disadvantaged, weak. In a transport sense, 'being on the outskirts' has a number of (mostly) negative effects which usually amount to:

- high costs for the transport of goods and people towards the central area
- increased travel times
- unreliable travel times
- low frequencies
- low levels of service
- restrictions and various 'obstacles' to overcome in order to reach the 'central' locations (e.g. restrictions in transit when crossing borders, or adverse weather, or bad infrastructure, etc)
- many transfers, and so on.

The above equally apply to regions within a country or whole countries within a wider continent or region of a continent.

Although in most cases peripherality is discussed in relation to the transport and communication isolation from the centre, the real issue is to understand the mechanisms that apply in the so called isolated or peripheral regions and cause the well known problems of weak economic growth, social dissatisfaction, temporal instability, etc. It is also important therefore, to understand how improved transport connections will help ease these problems and stimulate economic growth. The purpose of this paper is to examine the relationship between overcoming isolation, transport and development by raising issues and providing some answers adopting a holistic approach i.e. viewing the problem from all its angles and not only transport.

## 16.2 The Gap Between Social and Economic Dynamics in Isolated Regions

The central premise of this paper is that the real problem of the so called 'peripheral' or isolated regions is one of initiating and sustaining economic activities that match the society's expectations and thus keep a balance between social and economic development that will ensure stability in economic, political, and social terms.

Historically, peripheral regions have developed as a more or less closed system in which the society grew in harmony with the economy. This meant that people's expectations for mobility, information, material goods, and quality of life in general was basically in line with the economy's ability to produce the required goods and services.

As technology progressed and information flows increased, especially after the relatively recent advances in communication and information transfer, the society's expectations grew more quickly than the economy and the production system could sustain. As the political and economic integration between the European states progressed, what the local economy could not provide was provided by importing goods and services from other economies thus starting the well known circle of imbalance in external trade, inflation, unemployment, and so on. Thus the real problem of peripheral regions seems to me to be simply one of closing the 'gap' between what society expects and what the (local) economy can provide. The real problems have to be looked upon as problems of how to foster sustainable growth that will meet society's expectations and thus reverse the negative circle of: low development - few opportunities - draining of human and material resources - further decline in growth and development opportunities.

The above circle of interaction between economy - society - and development is what today defines the specific situation of peripheral regions. An example of this circle is given in Figure 16.1. As a result of this circle of interaction, and according to local influencing factors which vary from area to area, there are a number of resulting problems and issues which are depicted in the middle squares of Figure 16.1. By taking space and location, i.e. regional and territorial planning issues as the main point of reference (as indeed this is the underlining issue in this paper), one can mention a host of problems faced by peripheral areas and countries today. These are, from the societal and demographic point of view: high demand for transport, environmental degradation, low utilization of natural resources, demographic decline, etc., and from the economic point of view: insufficient infrastructure, stagnation of agriculture, high costs of installation for industry, low productivity, high transport costs, etc.

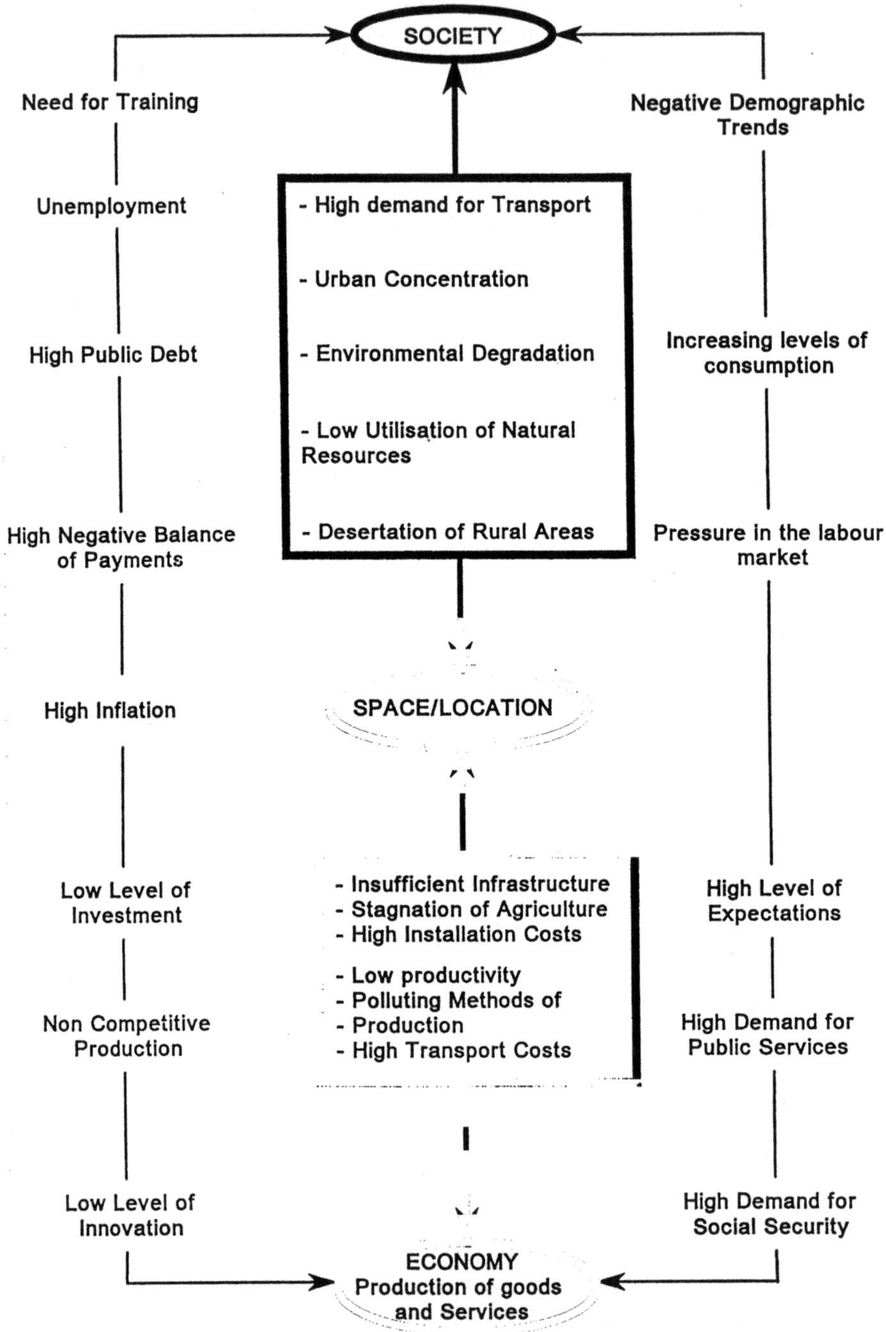


**Fig. 16.1.** The 'Circle' of social and economic dynamics in peripheral countries

The traditional reaction of governments to the above circle of interaction is to intervene in various ways in order to satisfy demand in the short or medium term (for obvious political reasons) thus, in most cases, creating a further imbalance to the situation and a further widening of the 'gap'. A long term, stable political strategy that recognises the nature of the problem and follows a consistent approach to its alleviation, is the first prerequisite for reversing the negative spiral of economic decline in peripheral regions. In this sense one could also state that the problems discussed here, are also political.

## 16.3 The Role of Transport and Communications in Fostering Development

The problems related to peripheral regions, require a number of long term measures and actions the success of which depends wholly, partly, or not at all on the existence of good transport connections with the central regions. Looking at the major actions and how much they depend on transport produces the picture of Table 16.1.

**Table 16.1.** The role of transport in fostering development

| | BASIC MEANS and... | ...THE ROLE OF TRANSPORT |
|---|---|---|
| 1. | Attract Capital Resources | Nil |
| 2. | Improve - Motivate Human Resources | Nil |
| 3. | Exploit Natural Resources | Outstanding |
| 4. | Exploit Cultural Heritage | Outstanding |
| 5. | Provide Raw Materials | Outstanding |
| 6. | Provide Access to Markets | Outstanding |
| 7. | Adopt Technological Innovation | Nil |
| 8. | Promote Institutional Change | Nil |
| 9. | Provide Political and Economic Stability | Nil |
| 10. | (many) Others | (almost) Nil |

From this indicative list of 10 major actions to foster economic development, the role and importance of the provision of adequate accessibility and good transport infrastructure is shown as outstanding in four of them. For the others this role is nil or not very significant. The main 'message' that comes from Table 16.1, is therefore that although Transport is a very important factor and perhaps a precondition for development, it is not but one of a variety of conditions and measures necessary.

Having said the above, what exactly then is the role of transport? Perhaps the most appropriate answer was given in a recent book (Giannopoulos and Gillespie, 1993) on the effects of transport and communications innovation in the building of the Europe of 2020 where it is noted that: "few changes are having a greater impact on the ability of firms and countries to compete in global markets than the recent and on-going revolution in telecommunications and transport. The new capabilities of information processing and transmission, as well as the enhanced mobility of people and the movement of freight, are profoundly altering features upon which the competitiveness of firms and the comparative advantages of regions depend".

The key forces generating a new industrial and spatial structuring are embodied in the radical technological changes currently under way in the telecommunications and transport industries. Communication and transport networks can be regarded as the 'carriers', in both the literal and symbolic sense, of new systems of industrial and spatial organization.

The idea of communications and transport as the carriers of new industrial and spatial forms is of course not new. Many commentators have drawn attention to the historical association between advances in transport and/or communications technologies and changes in the nature of society, changes in the way the economy is organized, and changes in spatial structure and organization. In one sense, the very existence of the city can be understood as the spatial response to the severe limitations upon the movement of people, goods, and information which prevailed before and during the early stages of, the industrial era. As some researchers have put it, "to avoid transportation, mankind invented the city".

Adopting a historical perspective, it is evident that advances in transport and communication have been clearly related to the establishment of a series of major economic, social and spatial developments in every country. We purposely do not use the term 'caused'; the complex question of the nature and direction of causality is the relationship between transport as well as technological innovation in transport, and society.

There are a number of points which are worth mentioning concerning this relationship:

1. This relationship primarily manifests itself in the impact it has on the spatial structure and organization forms of the system of production. Specific organizational forms of production, for example Just-in-Time production, require both advanced telecommunication infrastructures as well as modern, efficient and reliable transport systems. The lack of one of these two infrastructure systems would inevitably lead to inefficiency and to difficulties in sustaining this specific spatial-organizational form of production.

2. The interaction between transport, and development (as expressed by many attributing factors such as the emergence of new organizational forms of production and spatial trajectories) can be regarded as a fundamental and crucial 'bridge' to capture the linkages between communication possibilities in general and spatial and economic dynamics, but this should by no means be taken as a straightforward, linear nor uni-directional relationship. At best it can

be regarded as a circular set of interconnections, making the definition of the original causes of change not easy to define. To present the argument in a linear narrative way, the circle must be 'broken into' at some point. Although we have chosen here to break into the circle at the point of the problems for peripheral regions, reflecting the context within which this paper is written, we contend that other starting points would be equally valid, and indeed may be more useful in understanding the nature of contemporary transformations between transport and development.

3. When considering the issues of peripherality and development, one should adopt a two level approach, a micro and a macro level. The micro level refers to the firm, i.e. to change in the intra- and inter-corporate organizational structure and in the spatial organization of production. These changes could manifest themselves through a new division of labour. The macro level relates to the industrial and economic system as a whole and underlines changes in universal industrial behavioral rules, in regulatory regimes, and to the spatial structure around which industrial systems are organized.

4. A clear characteristic of the peripherality perspective is that no single, unique trajectory of change in the industrial and spatial structures can be identified for the future as solely or mostly related to transport. The development of transport and telecommunication technologies are of course very significant, but these in turn depend on the development trajectories of some other crucial elements, including technological innovation, the diffusion and application of technologies throughout the industrial system, and the institutional framework within which diffusion takes place - concerning for example the structure of markets and how they are regulated, standards issues, etc.

These elements can follow a number of different development paths, each of them leading to a different pattern of usage of transport and telecommunication technologies and consequently to the constitution of different industrial and spatial structure scenarios.

5. From the perspective of regional areas, the effects of transport and communication, in production structure, organization and technology have both positive and negative implications. New organizational models might contribute to decentralization of decision-making from centres of markets and administration while some growth industries which are locationally more flexible and not so tied to central locations are likely to be attracted. In contrast, face-to-face contacts and utilization of special services, which are more easily accessible in 'central' surroundings, play an increasingly important role in modern economic activities. These elements emphasize the strategic role of transport and communications in the adaptation of regional economies to a wider national and international economy.

So, in summary, the relationship between transport and development is an established and well accepted fact, which nevertheless must be seen as a vital but otherwise not exclusive factor of success.

## 16.4 The Relationship Between 'Peripherality' and Development

There are several ways of looking at this relationship. From the 'statistical' point of view, i.e. from current experience, one can easily note that 'being on the periphery' usually means low development, depopulation, few opportunities for economic activity, low productivity of the production forces, and all the other problems depicted in diagrammatic form in Figure 16.1. Deviations from this picture do, of course, exist and to this point one must refer to the comments about the diversity of the existing internal conditions that are made later on in this section.

The other way of looking at this relationship is from the perspective of a more analytical approach that aims at identifying the positive as well as the negative sides of this relationship and the prospects for future utilization of the positive ones. From this analytical viewpoint we will concentrate a bit further as it is generally felt (and we believe too) that 'peripherality' should not necessarily mean a low level of development.

Traditionally, the socio-economic and spatial development of a region depended on the basic factors of production i.e labour, land, and capital. In today's peripheral regions in Western Europe and elsewhere, labour and capital as factors of production are scarce. Contrastingly, the role of the land, especially in the countryside, as a compensation area for urbanized society has become more prominent. It provides milieux for housing and industry (if appropriate incentives are given to compensate mainly for the lack of good transport and communications links), as well as leisure and tourism and can be utilized as a reserve of both natural and cultural landscape. Consequently, the role of space as the basic resource of 'regional' areas is accentuated. Competition for it intensifies in most accessible regions whereas there is a surplus of it in most remote areas. This is the first and perhaps most important impact of the provision of transport on the 'development' of peripheral areas.

Regarding now the other two factors of production, namely labour and capital, here too the position of a peripheral region should not necessarily be one of weakness in the future. The increasing pressure to decentralise economic activities on the grounds of environmental protection of congested central locations and the political pressures to decentralise decision making processes, are the most promising forces that, in the future, both labour and capital resources will be increasingly attracted back towards peripheral regions and countries.

The extent of the opportunities that will arise, will depend on the 'local conditions' in each case since peripheral areas in Europe today are internally differentiated to the extent that it is increasingly difficult to analyze them in terms of 'averages'. There is extensive variation which concerns both current structural characteristics and ongoing development policies and processes. Take for example the differences between the southernmost and northernmost 'isolated' regions in Western Europe. The former are still characterized by the rural exodus syndrome and by deficiencies in the provision of basic services. In contrast, the very thinly populated regions in the north are fairly well equipped with up-to-date infrastructure networks, but they face the intricate problem of how their small and scattered centres could develop into dynamic nodes in a modern international economy. As a consequence, the need to adjust strategies according to local circumstances is obvious, both in the provision of services and in the restructuring of economic activities.

One point that should not be missed concerns the potential conflict with the ongoing infrastructure development policies for the peripheral areas, which are to an increasing degree, based on the priorities of the urbanized society and, more importantly, as seen from an international 'central' perspective. The emphasis placed on demand and competition is difficult to bring into accord with the requirements especially of rural regions. Advances in network infrastructure technologies offer some new options, and solutions are primarily sought by choices in transport and communications policies, but these must be parallel to other most important decisions about the nature and function of the markets in these areas, and the other cultural and educational needs of the local populations.

Relative to the subject of peripherality and development, we would say that there are some priorities in policies which are worth emphasizing:

- First, the modernization of regional transport and communication networks is a necessary precondition for the utilization of their potential for development but not sufficient. There is a need for special-purpose financial support programs and also setting the modernization of such networks as an important target in regulatory practices concerning network service operators in a more competitive environment.

- Second, upgrading demand is an essential element in development strategies in regional areas. Improvements in social infrastructure, especially in education and cultural life as well as in the economy of the region are absolutely necessary to be effected with measures employed in the context of an integrative program, which presupposes co-operation between organizations at different levels.

- Third, public service obligations should not be abandoned as a point of reference in policies, although in practice relevant measures (e.g. subsidized tariffs, differentiated service charges, etc) have to be studied in detail and according to resources, technologies, and specific circumstances.

- Finally, public procurement and R&D support with special emphasis for regional communications and transport applications are needed, because there is, generally speaking, a 'central' bias in the technical development in network infrastructure and services.

## 16.5 Some Restraining Factors

The debate on transportation, communication and mobility and their relationship to the development of the so called isolated and peripheral areas can not avoid serious questions about the environment, social justice and emancipation as mobility itself can become part and parcel of a process of uneven development and of consolidating asymmetrical development relationships. Raising questions of integrated policies (i.e. taking into account social and other considerations) may be seen to belong outside the scope of this paper and indeed to an outmoded debate, but absence of these considerations would render the analysis incomplete and would fail to explain why certain peripheral regions, even after construction of modern transport and communications networks are still caught in the doldrums of persistent deprivation and perpetual restructuring.

The interest in infrastructure over the past few years has by no means excluded the very important issues of the environment and more recently the notion of 'sustainable' mobility and development. Infrastructure systems are also becoming more and more of an issue because of shifts in user requirements and technological innovation. In addition to improved means of transport (vehicles and vessels), upgraded regulation of traffic and transport flows, improved utilization of network capacity and better alignment of the various transport modes and infrastructures, it is argued that the accelerated process of computerization of our society should go together with a new way of thinking in order to find creative solutions based on fundamentally new approaches to the problems in hand. What the present situation demands from the modernization of the physical infrastructure is not only repair, replacement and optimization of existing systems, but technological modernization of entire systems, including their institutional and social components.

Many of the developments in communication and information technology and in logistics will emerge in different ways, at different times and at different speeds. It is therefore difficult and not advisable to control these developments in a top-down manner in order to foster development of peripheral areas. This will cause inflexibility and inefficiency. However, the proliferation of all kinds of systems may also be unattractive, and some kind of assistance in the required standardization, or help in achieving horizontal and vertical cooperation is valuable. This is an important task for governmental organizations.

All over the world, deregulation and privatization of public facilities is emerging. This will create possibilities for increased efficiency through increased competition amongst companies and countries. For peripheral areas, there is a danger here in that monopolistic tendencies may occur, and governments should carefully observe what is happening in order to take appropriate actions. This means that there is an important need to monitor the developments mentioned above. Many of the statistics that we use are not well fitted to achieving this task. Especially in the field of communications and the organization of logistics, some efficient form a data collection has to be developed.

This process of monopolistic forms of economic activities in peripheral areas has only just begun. The so called 'healthy' competition conditions for the development of peripheral regions must be preserved and there are many obvious safeguards, such as social, technical, organizational, financial and legal barriers, to be established.

Another important point concerns the economic justification and the financing method for the necessary development of infrastructure. In order to achieve optimal usage of scarce resources, all of the costs which society has to pay, in order to facilitate activities, have to be taken into account, and ideally be met by the users of these facilities. In many cases this is not easy and at present, many costs, especially those related to environmental damage, are generated but not compensated for. Of course, the internalization of external costs - also relevant in areas other than transportation and logistics - could lead to significant changes in logistical decision making, and therefore is of utmost importance.

## 16.6 The Perspective of the Aegean Islands in Greece

Greece has traditionally been a peripheral country. Its development during the 150 years of its existence as a free country in modern history, is marked by the deficiencies and problems of 'peripherality' as defined in this paper. Especially after the second world war, the process of development, which culminated with Greece's entry to the EEC in 1981, is a case study in which one can find all the elements of the interaction between transport and development, and the 'gap' between societal and economic development that were outlined previously. As a result, Greece still has one of the lowest GDP per capita in Western Europe, and serious structural problems in her economy, and this despite 10 years of generous economic assistance from the European Union.

The islands of the Aegean, and especially those of Northern Aegean, are an example of a 'peripheral' region within a 'peripheral' country. With the exception of the larger touristically developed islands (such as Rhodes, Crete, Mykonos) the majority of these islands (and there are almost 300 inhabited islands today) suffer from increased isolation from the Greek mainland and show all the characteristics of such areas that were discussed above.

Several programmes and special measures have been discussed for the development of this region and all of them have as a central point the improvement of the transport and communication links especially during the winter months. Not disregarding our main points concerning the need to take a holistic approach (i.e. measures for the local economy, etc.) along the lines discussed earlier in this work, we can state as the most important suggestions that have been put forward in the transport field as follows:

- Creation of a basic network of transportation focal points (hubs) where air and sea connections to the mainland and abroad will be improved and upgraded. Islands which have been proposed as these hubs are:
  * Lesvos (Mytilene)
  * Kos
  * Mykonos
  * Samos
  * Rhodes
  * Heraklion
  * Skiathos

**Fig. 16.2.** Major transhipment nodes in Greece

- Development of regular, frequent connections by sea and air with focal points as the other islands in their 'vicinity'. The whole hub and spoke system will be formulated through an appropriate Master Plan study.

- Development of a network of heliports in the lesser populated and isolated islands in order to provide the local population with emergency transport services, thus increasing the reliability of the transport network for these islands.

- Introduction of telematics applications in the whole transport and communications infrastructure of the Aegean. This means applications such as traffic and travel information for all means of transportation, improved data and information transfer networks between ports and airports in the area, telemedical services, teleworking and teleconferencing, advance weather information, etc.

- Stimulation and support of certain (minimal) transport service operations for the more remote and isolated islands where normal (competitive) services cannot be maintained.

A final and perhaps most important role for the whole Aegean area, will come from the improved position of Greece as a 'central' country in terms of geographic position and transportation services after the recent events in the area. More specifically the events in Europe the last four to five years are producing a dramatic change in Greece's position. The notion of peripherality has been developed from the economic and political reality of the 45 post war years when Greece was at the edge of one of the two 'blocks' of sociopolitical systems i.e. Western Europe. For 45 years she had political and economic ties predominantly with her partners in the West, while the cold war politics prevented any meaningful relationship with the neighbouring countries of Eastern Europe. As a result all the commercial and cultural (including tourism) exchanges were between Greece and the Western European countries thus making Greece a 'peripheral' country in the true meaning of the word. Transport flows had to transverse great distances in transit through many countries in order to reach final destinations in central and Western Europe and as a result Greek products and raw materials had an added cost of transport, unreliable travel times, and insufficient transport capacity available during the months of peak demand, and all the other poor transport characteristics of 'peripheral' regions.

Today, this situation is dramatically changing. The opening up of the former Eastern European economies, with the resulting increase in commercial and other interactions between these countries and the rest of Europe, has decidedly shifted the centre of European 'gravity' eastward. At the same time the other traditional pole of economic and cultural activity, the Arab world, emerges after years of internal conflict, to be a source of future dynamism that will look towards Europe as its natural field of interaction.

Greece, and the Aegean more particularly, finds itself therefore in an increasingly central position at the point of interaction of three worlds. As a result

Greece can start to play a more 'central' role in providing infrastructure and services that will serve the flows of goods and passengers among these three groups of socioeconomic entities. In this context, the Aegean can play a 'central' role in providing transportation services to the rest of the Eastern Mediterranean and the Black sea. This picture is depicted in Figure 16.2 and consists of the following four main points:

a. A central trans-shipment port should be developed in the central Aegean in one of the islands where this is environmentally acceptable. This could be the island of Makronisos near Attica or also other similar points.

b. The upgrading of the port of Thessaloniki and its road and rail connections to the rest of the Balkan countries, so that it becomes a central transit node for the whole of the Balkan area.

c. The upgrading of the port of Alexandroupolis, and its orientation towards serving the flows from Bulgaria and Rumania, and

d. the reopening of the ferry boat service from Volos to the Middle East and the extension of the port of Volos to become a major combined transport terminal in the Eastern Mediterranean.

## 16.7 Conclusions

The central premise of this paper is that the real problem of the so called 'peripheral' regions is one of initiating and sustaining economic activities that match the society's expectations and thus keep a balance between social and economic development that will ensure stability, in economic, political, and social terms.

Provision of good transport and communication facilities is an absolutely necessary condition for the initiation and sustainment of economic development but it is by no means the provision of transport alone that will foster development. In fact the interaction between transport, and development can be regarded as a fundamental and crucial 'bridge' to capture the linkages between communication possibilities in general and spatial and economic dynamics, but this by no means should be taken as a straightforward, linear nor uni-directional relationship. At best it can be regarded as a circular set of interconnections, making the definition of the original causes of changes not easy to define.

From the perspective of the so called isolated or peripheral areas, the effects of transport and communications, in the production structure, organization and technology have both positive and negative implications. New organizational models might contribute to decentralization of decision-making from centres of markets and administration while some growth industries which are locationally rather flexible and not very tied up to central locations are likely to be attracted.

In contrast, face-to-face contacts and utilization of special services, which are more easily accessible in 'central' surroundings, play an increasingly important role in modern economic activities. These elements emphasize the strategic role of transport and communications in the adaptation of regional economies into a wider national and international economy.

Concerning now the relation between 'peripherality' and development it can be said that by looking at these issues from a purely 'statistical' point of view, one can note that indeed 'being in the periphery' usually means low development, depopulation, few opportunities for economic activity, low productivity of the production forces, and other problems which at the end amount to what we usually call low level of development. However, when looking at this relationship from the perspective of a more analytical approach one finds some strong positive points in 'being peripheral' which, if properly utilized can lead to positive prospects for the peripheral areas. This concerns two of the three factors of production namely land, and capital which if properly manipulated will attract the third, i.e. labour. Capital can be attracted through a series of proper incentives and safeguards which have been tested as part of several comprehensive regional development policies in the past. The role of the land in peripheral countries, especially in the countryside, will also become more prominent as a compensation area for urbanized society. It can provide milieux for housing and industry as well as leisure and tourism and can be utilized as a reserve of both natural and cultural landscape. Consequently, the role of space as the basic resource of 'peripheral' areas is accentuated. Competition over it intensifies in most accessible regions whereas there is a surplus of it in most remote areas. This is a most important impact of the provision of transport on the 'development' of peripheral areas.

The extent of the opportunities for development of the peripheral regions, will finally depend on the 'local conditions' mainly the structural characteristics of the economies, and the ongoing development policies and processes. Transport can play a very important role, and must be provided as a precondition for development, but it must be one of a whole package of actions that must be taken.

As regards the area of the Aegean in Greece, a number of action lines are suggested but the point is also made that after the 'revolutionary' changes in Europe in the last five years, this area now finds itself in the middle of three 'worlds', the economically powerful western Europe and primarily the European Union to which she belongs, the emerging new economic entities of former Eastern European countries, and the Arab world who is now on the verge of a new peaceful era which undoubtedly will bring new prosperity and economic activity. Greece therefore must exploit her new 'central' position and develop the functions of a central area and of a 'bridge' linking these worlds. It is therefore presented with new opportunities for development which if developed further will transform the area from an 'isolated' and 'peripheral' part of Western Europe, to an economically viable 'central' area providing transportation services for the whole of the South, South Eastern Europe.

## References

Giannopoulos G, Gillespie A (1993) Transport and Communications Innovation in Europe. Belhaven Press, London

# Contributors

G. Boulougaris
University of Thessaloniki
Civil Engineering Dept.
Transport Research Unit
54006 Thessaloniki
Greece

R. Capello
Politecnico of Milano
Piazza Leonardo da Vinci
20133 Milano
Italy

H. Coccossis
Dept. of Environmental Studies
University of the Aegean
17 Karantoni Street
81 100 Mytilini
Greece

M.M. Fischer
Dept. of Economic and Social Geography
Vienna University of Economics and Business Administration
Augasse 2-6
1090 Vienna
Austria

P. Freudensprung
Dept. of Economic and Social Geography
Vienna University of Economics and Business Administration
Augasse 2-6
1090 Vienna
Austria

M.S. van Geenhuizen
Free University
Faculty of Economics
De Boelelaan 1105
1081 HV Amsterdam
The Netherlands

G.A. Giannopoulos
University of Thessaloniki
Civil Engineering Dept.
Transport Research Unit
54006 Thessaloniki
Greece

M. Giaoutzi
Dept. of Geography
National Technical University
Zographou Campus
Athens
Greece

F. Hiemstra
Free University
Dept. of Regional Economics
De Boelelaan 1105
1081 HV Amsterdam
The Netherlands

G. Kafkalas
Dept. of Urban and Regional Planning
Aristotle University
University Box 491
54006 Thessaloniki
Greece

S. Kostopoulou
Aristotle University
University Box 491
54006 Thessaloniki
Greece

K. van Montfort
Free University
Faculty of Economics
De Boelelaan 1105
1081 HV Amsterdam
The Netherlands

P. Nijkamp
Free University
Faculty of Economics
De Boelelaan 1105
1081 HV Amsterdam
The Netherlands

G. Pepping
Free University
Faculty of Economics
De Boelelaan 1105
1081 HV Amsterdam
The Netherlands

H.N. Psaraftis
Dept. of Naval Architecture and Marine Engineering
National Technical University
Zographou Campus
Athens Greece

Chr. Rammer
Dept. of Economic and Social Geography
Vienna University of Economics and Business Administration
Augasse 2-6
1090 Vienna
Austria

R. Ratti
Centre for Economic Research
Cantone Ticino
6501 Bellinzona
Switzerland

A. Reggiani
Department of Economics
University of Bologna
Piazza Scaravilli, 2
40126 Bologna
Italy

P. Rietveld
Free University
Dept. of Economics
De Boelelaan 1105
1081 HV Amsterdam
The Netherlands

R.-J. Simons
Free University
Dept. of Regional Economics
De Boelelaan 1105
1081 HV Amsterdam
The Netherlands

K. Spiess
147 Redfield Place
Syracuse, NY 13210
USA

A. Stratigea
Dept. of Geography
National Technical University
Zographou Campus
Athens
Greece

R Vickerman
CERTE
Cornwallis Building
The University
Canterbury, Kent CT2 7NF
England

R. Zwier
Free University
Dept. of Regional Economics
De Boelelaan 1105
1081 HV Amsterdam
The Netherlands

MIX
Papier aus verantwortungsvollen Quellen
Paper from responsible sources
FSC® C105338

If you have any concerns about our products,
you can contact us on
**ProductSafety@springernature.com**

In case Publisher is established outside the EU,
the EU authorized representative is:
**Springer Nature Customer Service Center GmbH**
**Europaplatz 3, 69115 Heidelberg, Germany**

Printed by Libri Plureos GmbH
in Hamburg, Germany